Green Energy and Environment

Edited by Eng Hwa Yap and Andrew Huey Ping Tan

Published in London, United Kingdom

IntechOpen

Supporting open minds since 2005

Green Energy and Environment
http://dx.doi.org/10.5772/intechopen.74000
Edited by Eng Hwa Yap and Andrew Huey Ping Tan

Contributors
Bilge Nazli Altay, Sylvain Cloutier, Martin Bolduc, Fritz Zaversky, Iñigo Les, Marcelino Sánchez, Benoît Valentin, Jean-Florian Brau, Frédéric Siros, Jonathon McGuire, Flavien Berard, Jiale Xie, Liuliu Wang, Xiaoying Chen, Pingping Yang, Fengkai Wu, Yuelong Huang, Laura Alejandra Sabogal-Moncada, Simon Montoya-Bedoya, Hader Vladimir Martínez-Tejada, Esteban Garcia-Tamayo, Fabrice Abunde Neba, Prince Agyemang, David Yahaya, Endene Che, Eyong Ndip, Razak Seidu, Andrea Majlingova, Martin Lieskovský, Maroš Sedliak, Marián Slamka

First published in London, United Kingdom, 2020 by IntechOpen
IntechOpen is the global imprint of INTECHOPEN LIMITED, registered in England and Wales, registration number: 11086078, 7th floor, 10 Lower Thames Street, London, EC3R 6AF, United Kingdom
Printed in Croatia

British Library Cataloguing-in-Publication Data
A catalogue record for this book is available from the British Library

Additional hard and PDF copies can be obtained from orders@intechopen.com

Green Energy and Environment
Edited by Eng Hwa Yap and Andrew Huey Ping Tan
p. cm.
Print ISBN 978-1-83880-571-5
Online ISBN 978-1-83880-572-2
eBook (PDF) ISBN 978-1-83880-573-9

Meet the editors

In the Faculty of Transdisciplinary Innovation, Eng Hwa Yap is the Acting Associate Dean: Teaching and Learning (Operations). He also leads the Bachelor of Technology and Innovation (BTi) as the Course Director. His research at the university focuses on curated multidisciplinary and mixed approaches of enquiry to understand and respond to complex problems surrounding technology, environmental sustainability, and future energy systems. In his research, Eng Hwa uses methods centering on systems thinking to investigate emerging and contemporary issues in technologies and their interaction with society, the environment, economics, and policy.

Andrew Huey Ping Tan has extensive engineering experience in both Motorola Solutions and Agilent Technologies, where he specializes in ensuring high quality and timely manufacture of industrial products for downstream customers. By implementing systems thinking and knowledge, life-cycle assessments, and technical improvisations, he specializes in product line setup and process improvements for new product developments. Andrew completed his PhD at the University of Nottingham where he developed a systemic measurement for the Water-Energy-Food Security Nexus in Malaysia using a transdisciplinary and systems approach. He also graduated with MEng (Hons) in Mechanical Engineering with First Class Honors from the same university.

Contents

Preface

Energy is a vital element in sustaining our modern society but the future of energy is volatile, uncertain, complex, and ambiguous; especially when facing a continuous drive to ensure a sustained and equitable access as well as mounting pressures to reduce its emissions. Traditional approaches in developing energy technologies have always been in isolation with distinct and unique contexts. However, we cannot afford to work in silos any longer. Future energy systems and their relationship with the society and the environment will have to be conceived, designed, developed, commissioned, and operated alongside and within contemporary geo-political, ethical, and socio-economic contexts. This has posed an unprecedented volatility, uncertainty, complexity, and ambiguity (VUCA), where systemic and holistic approaches are often warranted. This book aims to focus on the VUCA of addressing the future of energy and environment by considering contemporary issues and insights from diverse contexts, viewed as a system, and anchored upon emerging and smart energy technologies.

Green Energy and Environment contributes to this on-going discourse and the role these energy technologies play in the society's effort to address climate change, energy consumption, and greenhouse gas emissions in a holistic and inclusive manner, by considering green energy technologies' key roles along a critical path to decarbonization and responsible innovation.

Green Energy and Environment is a collection of timely efforts to address the complexity of developing energy technologies within these unprecedented contexts, each contributing to the main theme and the on-going discussion from diverse lenses and a holistic viewpoint. Topics covered in the book range from green electronics, techno-economic optimization of solar-only powered combined cycle power plants, a case study of hydrovoltaic materials for China, to circular economy of electro-chemical energy storage. Each chapter addresses an important aspect of energy technologies and serves as a vital building block towards constructing a timely and relevant body of knowledge on green energy and environment.

We hope you will find this book useful and the topics relevant. This book is a result of many months of hard work by all contributing authors, who have worked tirelessly and passionately. Without them, this work would not have seen the light of day. It is in this regard that we would like to acknowledge the professionalism and commitment from all chapter authors – thank you! We are also grateful to IntechOpen for this opportunity, and the Publishing Process Managers Lada Bozic and Danijela Pintur for their wisdom, friendship, encouragement, and assistance throughout the process of editing this book – thank you!

Eng Hwa Yap
University of Technology Sydney,
Australia

Andrew Huey Ping Tan
KDU University College,
Malaysia

Chapter 1

Sustainable Advanced Manufacturing of Printed Electronics: An Environmental Consideration

Bilge Nazli Altay, Martin Bolduc and Sylvain G. Cloutier

Abstract

Printing technologies have become a novel and disruptive innovation method of manufacturing electronic components to produce a diverse range of devices including photovoltaic cells, solar panels, energy harvesters, batteries, light sources, and sensors on really thin, lightweight, and flexible substrates. In traditional electronic manufacturing, a functional layer must be deposited, typically through a chemical vapor or physical vapor process for a copper layer for circuitry production. These subtractive techniques involve multiple production steps and use toxic etching chemicals to remove unwanted photoresist layers and metals. In printing, the same functional material can be selectively deposited only where it is needed on the substrate via plates or print heads. The process is additive and significantly reduces not only the number of manufacturing steps, but also the need for energy, time, consumables, as well as the waste. Thereby, printing has been in the focus for many applications as a green, efficient, energy-saving, environmentally friendly manufacturing method. This chapter presents a general vision on green energy resources and then details printed electronics that consolidates green energy and environment relative to traditional manufacturing system.

Keywords: additive manufacturing, printing, flexible electronics, functional inks, subtractive manufacturing

1. Green energy, environment, and electronics

Sustainable and renewable green energy and materials as an alternative to fossil fuels that take millions of years to be developed have been the most important challenge for all industries to secure the future energy demands, environment, and human health [1]. Burning fossil fuels for energy, production and transportation of fossil fuel-based materials, industrial/agricultural activities, as well as growing population yield greenhouse gasses (GHGs) that trap heat in the atmosphere [2]. The GHGs remain in the air for various amounts of time, from a few to thousands of years, causing global heating and drastic changes in climate [3, 4]. Therefore, innovations in all fields are critically important to reduce the GHGs, unsustainable energy and material usage, cost, toxic waste, and pollution which are the potential risks on human health and environment [5].

Worldwide energy consumption by source recorded to be an average of 18.4 trillion watts (TW) in 2018 [6]. **Figure 1** represents that the majority of the energy was based on fossil energy sources. For the future, the total consumption is projected to be 27.6 TW by 2050 and 43.0 TW by 2100 [7]. Researchers help formulating solutions to increase green energy production that comes from the natural sources such as solar, wind, ocean or tidal, hydropower, biomass, and geothermal energy. They are also called C-neutral sources [7]. Among these, solar energy is the largest source that enables more energy in an hour to the Earth than all of the energy consumed by humans in an entire year (if only this energy could be stored) [1]. Each energy sources have different potential to provide the projected power need. The theoretical delivery potentials of green sources in **Table 1** represents that using direct radiation from the sun is by far the only biggest source of energy [7].

The sun is a massive reactor where hydrogen atoms are fused into helium. The energy from this reaction is released into space in the form of radiation that creates electromagnetic energy—the entire range of light that exist (**Figure 2**). By using various technologies (solar panels and photovoltaics (PV)), the solar radiation can be turned into heat and electricity [8].

Part of the light radiated from the sun does not reach to the Earth due to various reasons [10]. Some portion for instance is reflected from the atmosphere back into the space, called *reflection of light*. Other portion is absorbed by the gasses and water

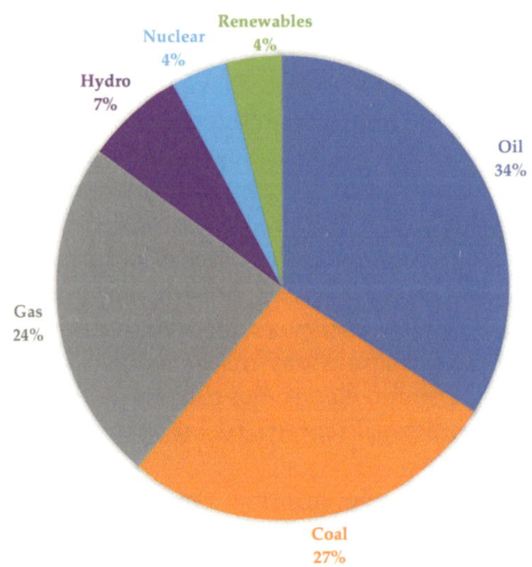

Figure 1.
Global energy consumption [6].

Green energy source	Theoretical potential (TW)
Solar energy	89,000 TW_p
Wind	1000 TW_m
Geothermal	44 TW_t
Hydropower	12 TW_m
Ocean tidal	2.4 TW_m

Subscripts denote mechanical, photonic, and thermal.

Table 1.
Energy delivery estimates of green energy sources.

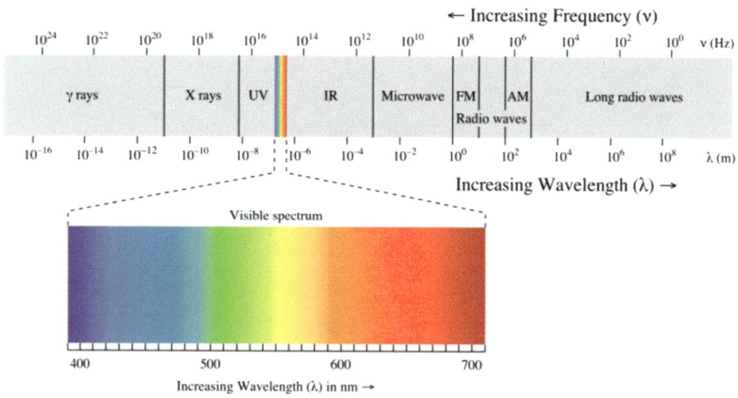

Figure 2.
Electromagnetic spectrum [9].

vapor molecules (O_2, O_3, H_2O, CO_2, etc.), called *absorption of light*. If the light comes on particles that are smaller than the wavelength of the radiation in the atmosphere, *Rayleigh scattering* occurs (mostly seen in gasses) and causes for instance the blue color of the sky. On the contrary, the particles larger than the wavelength of the radiation cause *Mie scattering* that happens due to aerosols and dust particles in the air. Therefore, the solar resource is broken down into three main components: (1) diffuse solar radiation (the light that are scattered), (2) direct beam solar radiation (the light that pass through the atmosphere), and (3) global solar radiation (the sum of (1) and (2)). The energy available from the sun that reaches the Earth, called *solar constant*, is considered to be 1367 W/m^2. However, due to the diffusion caused by the scatterings, usually 1000 W/m^2 (at 25°C, AM1.5G spectrum) is used to describe 1 day atmospheric condition for the standard test conditions of efficiency estimations of solar power [11]. The diffuse light is also known to limit the power generation efficiency of solar panels, alongside with the limited angle placement options for the rigid panel orientation (**Figure 3**), geographical location, time of the day, season, local landscape, and weather [12]

The basic working principle of solar panel is converting light energy directly into electricity through the photovoltaic effect. The panel is usually constructed by an n-type and p-type semiconductor material (silicon based in general) between the two metal conductor layers (**Figure 4**). The n-type semiconductor has extra electrons that carry negative charge, while the positive p-type semiconductor has

Figure 3.
Schematics of a conventional solar panel at different tilting angles.

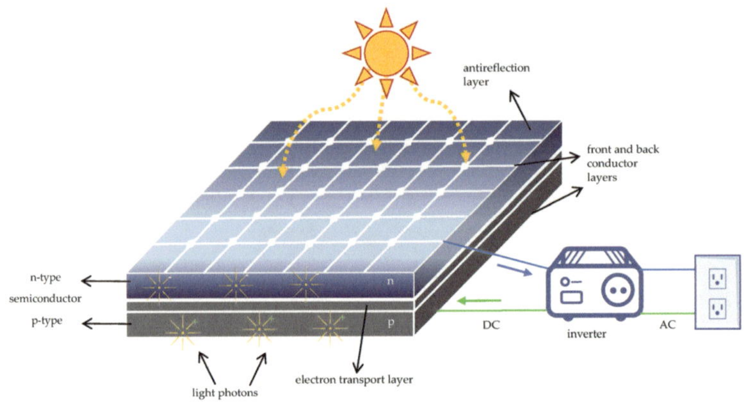

Figure 4.
Illustration of sunlight conversion to electricity.

missing electrons. When the light photons are absorbed, the extra electrons of the n-type get free (the so-called holes) and forced to travel in the electron transport layer by the top conductor. Meanwhile, the conductor on the bottom layer forces the missing positive electrons to travel in the transport layer. These moving electron–hole pairs induce the DC electric current formation that is converted into AC in the inverter unit of the solar system. In the case of solar heating (**Figure 5**), a panel of tubes heats up the water through the absorbed light energy and redistributes into the building for heating, air conditioning, and hot water usage.

Between the global energy need for electricity and heating, 10% is estimated to be for illumination purpose, while 90% is for the heat used to make products and to heat and cool buildings and homes and the energy used to drive motor vehicles [1]. One of the most critical factors is understanding not only how to produce the green energy, but also how to remanufacture, reduce, and reuse/recycle electronic products that use this energy. Thereby, as much as the effort goes into producing green energy from natural sources, the same effort is needed for electronic manufacturing since an electronic circuit is found in a surprising number of devices that we use in our daily life: from lighting to domestic/industrial appliances; from computers and its accessories to communication devices and cameras; from vehicle electronics to medical devices; or from the products that use displaying units, controlling apparatus, and switches to alarm systems and toys. The circuitry use is almost endless.

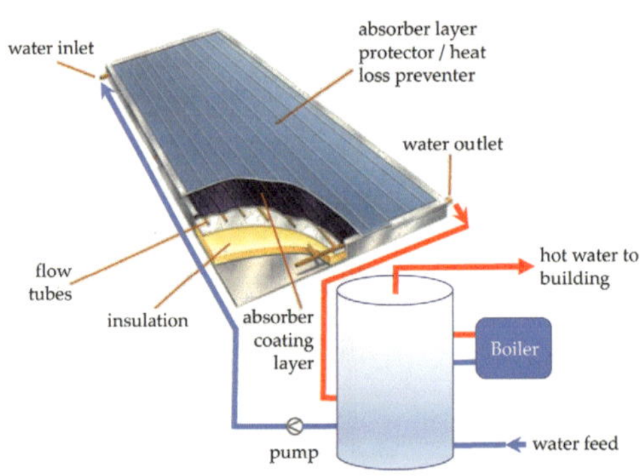

Figure 5.
Illustration of traditional solar water heating (edited) [13].

2. Traditional electronic manufacturing vs. printed electronics

2.1 Traditional electronic manufacturing

Traditional electronic manufacturing requires multiple production steps as illustrated in **Figure 6** [16]. First, a functional layer must be deposited on a substrate, typically through a chemical vapor or physical vapor process, for a copper layer production. The most common substrate used for the circuit board is a glass fiber-reinforced epoxy resin. Then, a photoresist layer is deposited on the substrate and experiences exposing, developing and curing processes. The next is the use of harsh etching chemicals to remove the photoresist layer and the unwanted metal that is not covered by the photoresist. The last step is striping the resist material and cleaning all the residues away. The entire manufacturing processes are highly time- and energy consuming, costly, and inherently wasteful. One main approach for green electronic manufacturing is called the three R's—"remanufacture, reduce, reuse/recycle"—focused on minimizing the use of energy, hazardous materials, toxic waste and pollution, and coolant consumption while machining, while promoting product take-back policies, the use of reusable/recyclable components, recycled feedstock in plastic parts, and lead-fee production [14].

2.2 Printed electronics manufacturing

Printing has received immense attention due to the additive nature of the manufacturing [15]. During printing, functional materials (or the so-called conductive/smart inks) can be patterned by selective deposition on where they are needed by the print heads in the case of digital printing, such as inkjet, 3D printing, and aerosol (**Figure 7**) or by the printing plates. Different functional materials are printed in a layer-on-layer manner, then followed by a curing process that forms necking between the pigment particles [16]. The additive approach and high production capability of the printing presses significantly reduce production cost, number of manufacturing steps, and the need for energy, time, and consumables, as well as offer great reduction in waste compared to traditional photolithography manufacturing. Printing allows sheet-to-sheet or roll-to-roll mass production; thereby electronics can be manufactured not only on rigid, but also on thin,

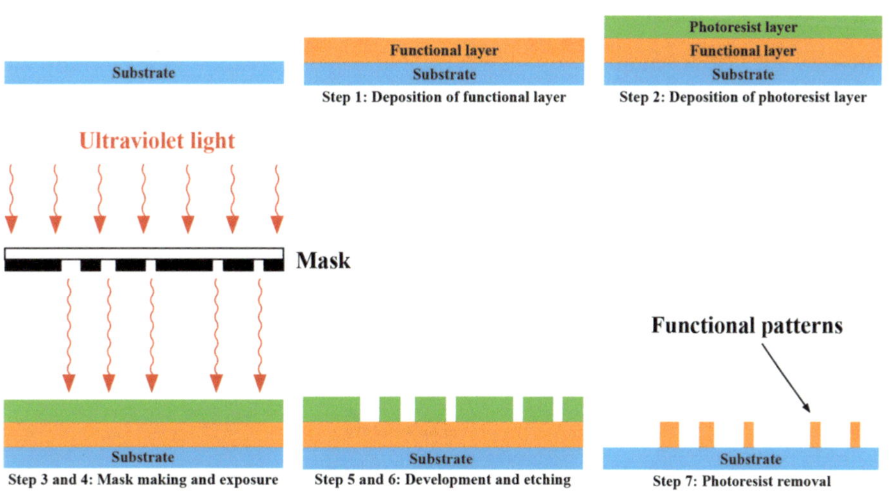

Figure 6.
Subtractive manufacturing steps [16].

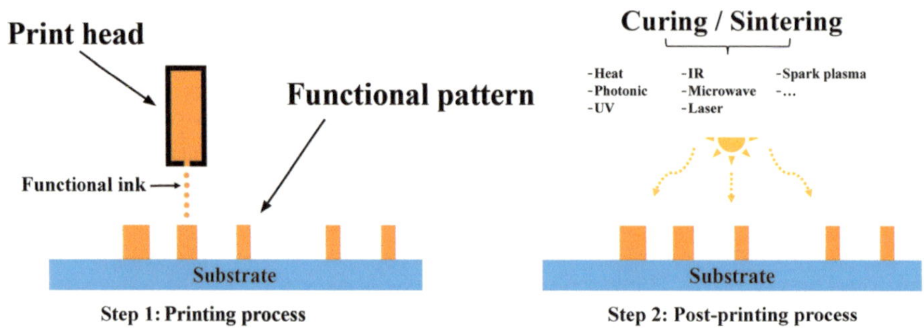

Figure 7.
Additive manufacturing steps [16].

lightweight, flexible, and large area substrates [17–22]. Printing technologies and the pluriformity of substrates open up launching brand new products that could have never existed before and realize bendable, rollable, wearable, or elastically stretchable devices. These printed electronics (PE) are environmentally friendly when compared to the traditional electronic methods. PEs are lightweight and not made with extremely harsh etching chemicals, and they do not occupy a massive amount of space in landfills. If the aim is to be green and sustainable, then printing technologies are certainly the future.

Printing techniques include conventional printing systems that require an intermediate printing plate to transfer a pattern (flexography, gravure, screen, and offset lithography) and nonimpact printing systems that print the pattern directly onto the substrates (digital and 3D) (**Figure 8**) [23]. However, nonprinting systems (liquid dispensing, aerosol) and coating systems (rod, blade, air knife metering) can also be used to dispense functional materials. The difference between the printing technologies originates from the ink characteristics (i.e., viscosity, rheology, surface tension), substrate types (i.e., papers, films, textiles), and printability properties (i.e., ink film thickness, resolution, speed, line quality) [24, 25]. **Table 2** shows some of the printed feature size capabilities of printing processes for PE applications.

In recent years, the advancements in digital inkjet printing technologies have shown great promises for printed electronics. Akin to more conventional additive manufacturing strategies such as using screen printing, digital inkjet printing has been rapidly and successfully applied for rapid prototyping, low-volume production, and hybrid integration of critical components for a wide range of optoelectronic applications including energy harvesting, wearables, and biomedical sensors [26, 27].

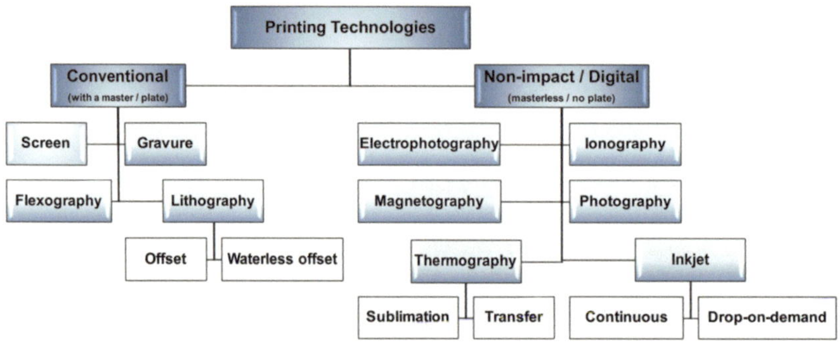

Figure 8.
Classification of printing technologies based on printing plate requirement [23].

Property	Screen	Flexo	Gravure	Inkjet
Viscosity (cP)	500–5000	50–500	100–1000	10–20
Minimum trace width (μ)	30–50	5–50	5–25	3–20
Minimum trace spacing (μ)	50–100	20–30	10–25	10–20
Ink film thickness (μ)	0.5–200	0.25–4	0.25–6	0.05–20

Table 2.
Common printed feature size of printing processes for PE applications.

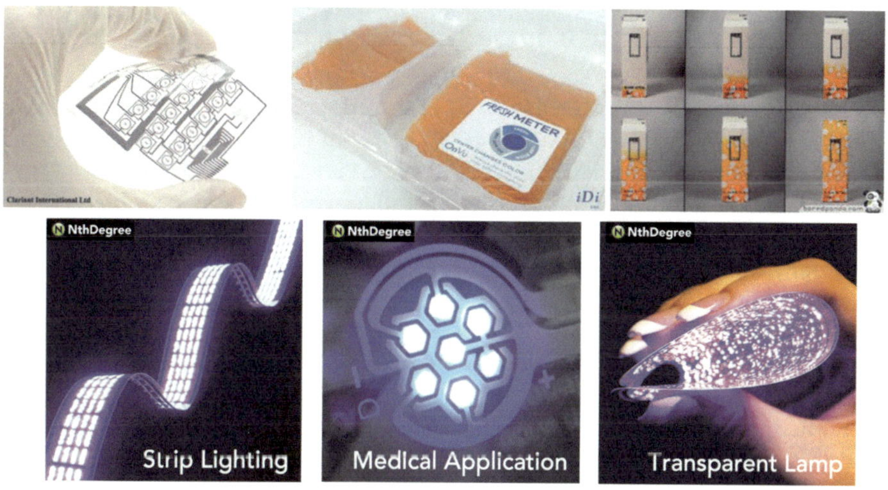

Figure 9.
Printed electronics samples: circuit printed with nano-silver ink [35], printed sensor that shows food spoilage [36], smart box packaging that enables end-of-life display by color change [37], and various lighting applications [38].

2.2.1 Functional inks

There are PE components that have been researched and fabricated using these printing techniques such as solar cells, displays, and transistors [28–31]. Similar to subtractive electronic manufacturing, PE components (**Figure 9**) require specific inks to provide functionalities like conductivity, resistivity, semi-conductivity, or color change by heat, light, moisture, pressure, or spoilage.

The materials listed in **Table 3** present common functional pigments used in ink formulations. A typical ink formulation includes binders, vehicles, and additives besides the pigments [32]. Binders are the chemicals binding formulation

Functionality	Pigment type
Conductors	Copper, silver, gold, carbon, aluminum, nickel, indium tin oxide, tin, graphene, graphene oxide, PEDOT:PSS, polyaniline, iron, graphite [39–58]
Semiconductors	Zinc oxide, silicon, zinc selenide, indium-gallium-zinc oxide, cadmium selenide, gallium arsenide, MALH [59–64]
Resistors	Aluminum oxide, hafnium dioxide, poly(4-vinylphenol), spin-on glass, parylene, solid electrolytes [65–72]

Table 3.
Functional pigment examples for PE applications [16].

ingredients to each other and to the substrate. Vehicle is the liquid portion of the formulation that carries the ink onto the substrates. Generally, ink formulations are classified based on the vehicle type, such as water-based, solvent-based, ultraviolet light/electron beam (UV/EB) based, or soy-based. Additives are used for supplementary properties, such as promoting stability, or preventing oxidation, flocculation, etc. [33]. In PE applications, binders and some of the additives act as an insulator and reduce the conductivity. There are studies that suggest binder-free formulations that reverse this impact and enhance conductivity [34].

In terms of solar energy harvesting, lighting and displays, and sensing applications, the research has pioneered new and better low-cost and printable optoelectronic materials and devices. Methylammonium lead halide (MALH) perovskites for instance (e.g., $CH3NH3PbX_3$, X either I, Cl or Br) have shown great potential due to their unique optoelectronic properties and the ability to replace P-N junctions for various applications including light-emitting diodes, solar cells, and photodetectors [64]. The power-conversion efficiency of photovoltaic devices has been reported to increase from 3.8% in 2009 [73] to 22.1% in 2016 [74, 75]. Such progress is largely attributed to improved processing and longer charge-carrier lifetimes directly related to increased material quality. Yet, fundamental challenges including low carrier mobilities still prevent the fabrication of large-area devices with performances competing with state-of-the-art technologies [76].

2.2.2 Substrates

The thermal and mechanical stability of substrates are critically important for the precise registration of functional ink layers upon each other to create PE components (i.e., electroluminescence lamp, capacitors, organic light-emitting diodes). Polymer films [77], papers [19, 78], flexible glass [79], textiles [80], and metal [81] have been given significant consideration as a substrate material. In PE applications, substrates either act as a base material to mechanically support electrical components such as circuit board [82], or as a top material for touch panels and display and lighting applications [83], or as an interlayer in batteries as separator membranes [84].

The quality and type of the substrate affect electrical, optical, mechanical, and magnetic properties of the functional ink layer [85] as well as economics [48]; therefore, its properties need to be engineered depending on the application. For instance, defects on the surface of the substrates may lead to pinholes and block electron flow in circuitry [77]. General requirements of substrates for PE include flexibility, transparency, surface smoothness, low thermal expansion, stiffness, heat resistance, low cost, thinness, and lightweight [86]. **Table 4** presents different properties of substrates having the same 100 μm thickness for flexible backplane applications (glass, plastic films [PEN and PI], and stainless steel) [22].

2.2.3 Post-printing process

In printing industry, drying is performed with an oven or via UV lamps; however, the inks used for PE applications require higher temperatures for densification or crystallization to function properly. Once the functional ink layer is printed, the post-printing process is needed to enable connectivity in pigment particles, so the printed ink layer can conduct electricity (**Figure** 7). The connectivity is accomplished by means of one or multiple processes: drying, curing, sintering, reactive chemistry transformation, and annealing. These post processes change ink structure by volatilizing vehicle of the ink and form interparticle necking between the pigment particles that grow particle grain to form continuous functional layer, while

Property	Unit	Glass	Plastics	Stainless steel
Weight	g/m^2	250	120	800
Safe bending radius	cm	40	4	4
R2R processable	—	Unlikely	Yes	Yes
Transparency	—	Yes	Yes/some	No
Maximum process temperature	°C	600	180–300	1000
Coefficient thermal expansion	ppm/°C	4	16	10
Elastic modulus	GPa	70	5	200
Permeable to oxygen, water vapor	—	No	Yes	No
Coefficient of hydrolytic expansion	ppm/RH%	None	11	None
Electrical conductivity	—	None	None	High
Thermal conductivity	W/m°C	1	0.1–0.2	16

Table 4.
Comparison of substrate properties for flexible backplane application.

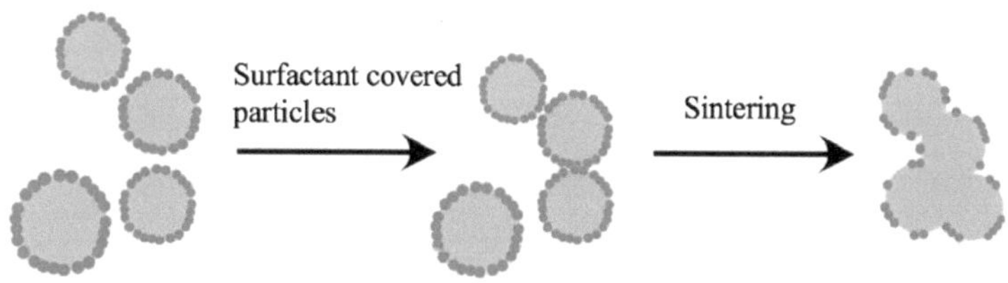

Figure 10.
Illustration of interparticle necking.

decomposing the binder (**Figure 10**) [87, 88]. The main types of post-printing methods are listed as microwave heating and electrical, spark plasma, laser, and photonic sintering [89].

The heat applied during curing volatilizes vehicle component of an ink formulation that allows functional pigments to contact each other. Sintering, on the other hand, is the process of pigment grain growth in the crystalline structure in the printed ink layer. However, the terms are used interchangeably. Sintering process requires optimization for each substrate as well as ink formulations since the chemical composition of the ink, particle size, shape and distribution, or degree of agglomeration varies. Different sintering parameters such as temperature, energy, time, or the atmosphere (ambient vs. inert) also cause variation in the performance of the same material. **Figure 11** shows an example of the effect of photonic energy variation on the sheet resistance of printed nickel ink [16]. As the energy applied increases, the sheet resistance decreases.

Photonic sintering has attracted great attention within the main types of sintering methods due to the instant heating applied during exposure, followed by an instant cooling, which is advantageous especially for the substrates with low glass transition temperatures. A pulse light from a xenon gas-filled flash lamp heats the functional ink layer in milliseconds beyond the maximum working temperature of the substrate. Then, the heat is removed rapidly in the interface of the substrate via conduction thanks to the thermal mass of the substrate and prevents structural degradation [90]. Three transient sintering conditions that are essential for an optimum photonic processing have been reported:

ink film thickness < substrate thickness
pulse of photonic light duration < thermal equilibration time of substrate
thermal equilibration time of ink film < pulse of photonic light duration

The thermal equilibration time of materials (τ_i) (s) is provided in Eq. (1), where K_i is the thermal conductivity (W/m K), ρ_i is the density (kg/m^3), c_{pi} is the specific heat (W s/kg K), and x_i is the thickness ink layer (m):

$$\tau_i = c_{pi}\rho_i x_i^2/4K_i \quad (1)$$

The thermal properties in **Table 5**, three transient conditions in **Table 6**, and the thermal profile presented in **Figure 12** exemplify photonic sintering of a 36-μm-thick nickel ink film printed on a 250-μm-thick solid bleached sulfate (SBS) paperboard that is processed at ~5 J/m^2 photonic energy that provides 12 Ω/□ of sheet resistance [16]. A millisecond of two overlapped light pulse heats the surface of printed nickel ink to a temperature between 350 and 500°C. Thanks to the rapid cooling, the temperature in the interface, 20 μm depth of the paperboard, reaches only 200–320°C. Although the ignition temperature of paper is 233°C, the heat

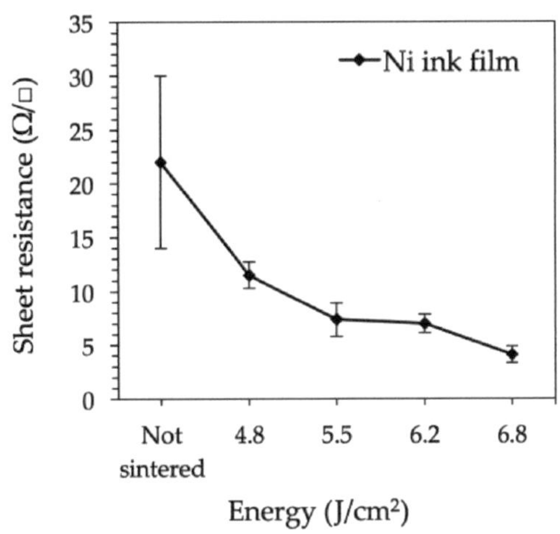

Figure 11.
Sheet resistance of printed Ni ink.

Factors	c_{pi}	ρ_i	x_i	K_is	τ_i
Nickel ink	440	8908	0.000036	90.9	1.4×10^{-5}
SBS	1400	900	0.000250	0.05	3.9×10^{-2}

Table 5.
Thermal properties of the ink and substrate.

Factors	
Ink film thickness < substrate thickness	36 μm < 250 μm
Pulse light time < substrate thermal equilibration time	0.001 s < 0.039 s
Ink thermal equilibration time < pulse light time	0.000014 s < 0.001 s

Table 6.
Three transient conditions.

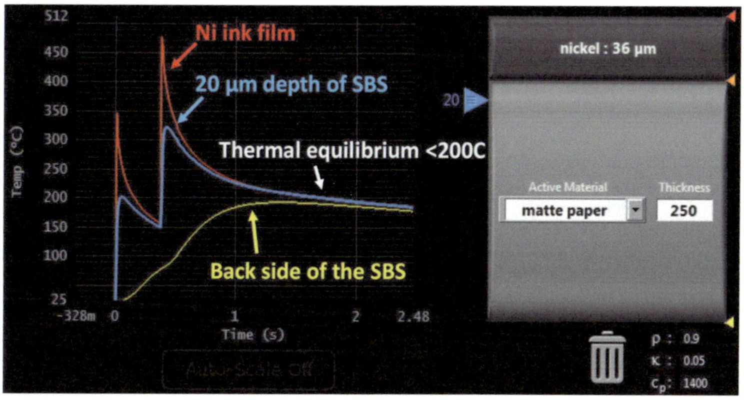

Figure 12.
Thermal profile simulation of nickel ink on a SBS paperboard.

equilibrates below 200°C in less than two milliseconds, which is too short for substrate to observe any deformation. Photonic sintering enables significant reduction both in processing time and energy in comparison to conventional oven that uses high temperature and processing time ranging from several minutes to hours.

3. Emerging advanced electronics manufacturing

In today's world, technological innovations accelerate at a much faster rate than before due to the more networked environment, advanced computers, data analytics, artificial intelligence (AI) tools, Internet of Things (IoT), and the speed of connectivity considering the ubiquitous information and communication technologies through Internet access, cloud computing, and smartphones [5, 91]. The inno vations in the printed electronics area derive mostly from flexible and conformable disruptive device designs and structures (single or multilayer circuit constructions, sensors), formulation of materials (inks, substrates), and manufacturing process design (printing, post-printing, assembly). The performance of printed devices is mainly dependent on the complex ink formulation, adhesion, and the interactions between the inks and substrates to produce materials that can withstand post-printing, assembly, and environmental processes [48, 92, 93]. Therefore, most material providers follow closed innovation model and keep proprietary rights for their complex material formulation, processes, and methods to stabilize their position in the market [5].

The most common issues with the traditional screen-printed circuit manufacturing market are the limitation of printing finely spaced traces (**Figure 13 (a)**), the printing process that requires new platemaking phase at each design and client change (**Figure 13(b-c)**), and the usage of high ink amount during printing which generates large amount of waste materials that has fairly complex disposal handling process (**Figure 13(d)**). In contrast, digitally printed electronic circuit manufacturing (**Figure 13(e)**) allows instantaneous design modification by simply changing the Gerber design file as well as dramatically reduces the material consumption [94]. The digital inkjet system is a rapidly emerging technology that could be in-lined to a hybrid automated component assembly pick-and-place robot systems (**Figure 13(f)**) where large-scale advanced manufacturing strategies can be explored as a potential way to reach seamless manufacturing of high volumes and open entirely new markets. However, the inkjet printing of functional inks requires

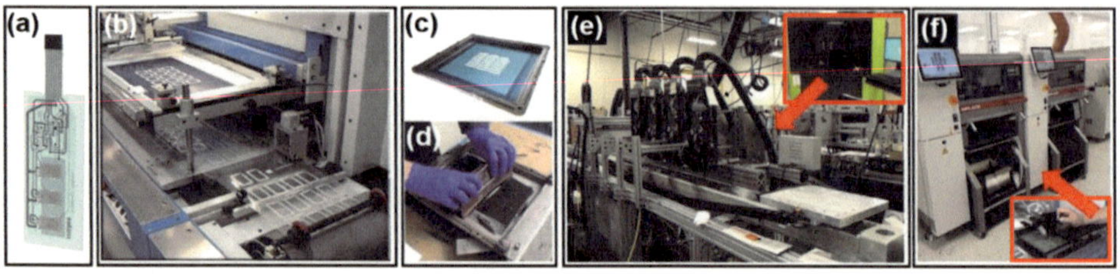

Figure 13.
(a) Commercial flexible hybrid electronic product that cannot be manufactured using screen printing (courtesies of Molex), (b) typical industrial grade screen-printing machine currently used for flexible circuit board manufacturing, (c) conventional screen stencil currently used for electronic board manufacturing, (d) typical screen printing application that uses tremendous amount of expensive functional inks, (e) industrial grade high-throughput digital inkjet printer vs. research grade in sectors like the graphic and consumer packaging industries (courtesy of Fujifilm), and (f) industrial grade fully automated pick-and-place robot for the assembly of electronic components of the circuit board (courtesies of C2MI and Varitron).

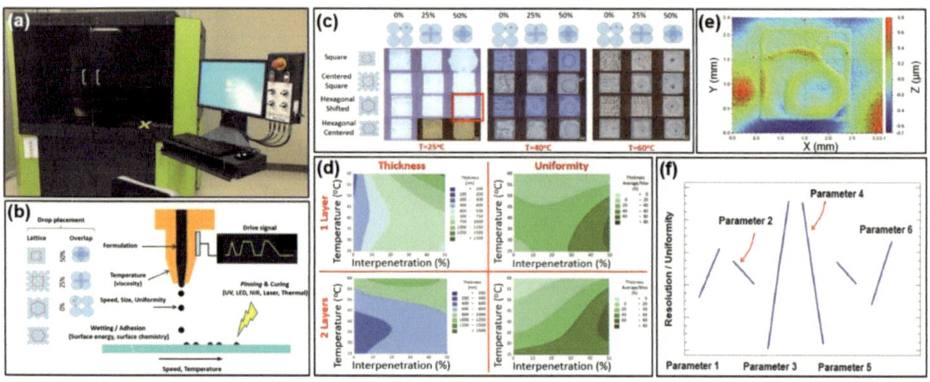

Figure 14.
Digital inkjet printing process mapping and optimization. (a) Research grade Ceradrop digital inkjet printing system at ÉTS; (b) precise control of a wide range of critical jetting parameters; (c) examples of inkjet-printed features using Ag ink on Kapton® for different printing lattices (drop placement configurations), drop interpenetration, and substrate temperatures; (d) example of one-layer and two-layer process mappings looking at the Ag features' thickness and uniformity for different substrate temperatures and drop interpenetrations in the hexagonal shifted lattice configuration; (e) typical high-resolution laser-scanning microscope image of the features obtained using a hexagonal shifted lattice with 50% interpenetration at 25°C substrate temperature; (f) example of a typical multivariate design-of-experiment analysis for process optimization (adapted from [94]).

a highly complex process optimization. **Figure 14** exemplifies a concrete process optimization done in ÉTS laboratory for only one silver ink formulation.

The massive push for intelligent cyber-physical systems associated with the Industry 4.0 and the IoT revolutions aims at taking better-informed decisions in real time, based on more complete and readily acquired sets of data. For the very reason, conformable printed sensors are deployed to collect data from places that are critical and difficult to access for energy, biomedical, transportations, manufacturing, smart building, or wearable electronics applications where better and cheaper flexible hybrid electronics circuits used as ubiquitous sensing elements would play a comprehensive role in ways that rigid devices cannot [95].

4. Conclusion

In this chapter, a general vision on energy sources and how an emerging field of printed electronics could consolidate green energy and environment is presented. An electronic circuit is found in a surprising number of devices that we use in our daily life. Manufacturing digitally printed electronic circuits is a sustainable method that dramatically reduces high energy consumption and toxic etching chemical usage relative to traditional electronic manufacturing. Advanced printed electronics is truly a transdisciplinary research and production landscape that benefit greatly from strongly intertwined interrelationships between multiple diverse complementary fields, including material formulation engineering, printable electronic devices architecturing, computational robotics and process automation, and AI and process optimization. It is important to adopt an agile mindset for a complete ecosystem to conduct transformative R&D for disruptive and advanced printed electronics manufacturing solutions.

Acknowledgements

This study was supported in part by the Scientific and Technological Research Council of Turkey (TUBITAK) under the 2214-A program.

Conflict of interest

There is no conflict of interest.

Author details

Bilge Nazli Altay[1,2,3*], Martin Bolduc[1,4] and Sylvain G. Cloutier[1]

1 Department of Electrical Engineering, École de Technologie SupÉrieure, Montréal, QC, Canada

2 Institute of Science and Technology, Marmara University, Istanbul, Turkey

3 Department of Chemical and Paper Engineering, Western Michigan University, MI, USA

4 Varitron, Saint-Hubert, QC, Canada

*Address all correspondence to: bilgenazli.altay@wmich.edu

References

[1] Lewis NS, Nocera DG. Powering the planet: Chemical challenges in solar energy utilization. PNAS. 2006;**103**(43): 15729-15735

[2] Environmental Protection Agency (EPA). Greenhouse Gas Emissions. Available from: https://www.epa.gov/ghgemissions/overview-greenhouse-gases [Accessed: 30 October 2019]

[3] NASA Global Climate Change. Vital Signs of the Planet. Solutions. Mitigation and Adaptation: Responding to Climate Change. https://climate.nasa.gov/solutions/adaptation-mitigation/ [Accessed: 30 October 2019]

[4] World Energy Council. World Energy Resources Report; 2016

[5] Lee SM, Trimi S. Innovation for creating smart future. Journal of Innovation & Knowledge. 2018;**3**:1-8

[6] BP. Statistical Review of World Energy; 2019

[7] Viswanathan B. Energy Sources: Fundamentals of Chemical Conversion Processes and Applications. Amsterdam, Netherlands: Elsevier B. V.; 2017. DOI: 10.1016/C2011-0-05048-2

[8] Tritt TM, Böttner H, Chen L. Thermoelectrics: Direct solar thermal energy conversion. MRS Bulletin. 2008; **33**(4):366-368

[9] File:EM spectrum.svg. Wikimedia Commons, the free media repository. 2015, November 22. Available from: https://commons.wikimedia.org/w/index.php?title=File:EM_spectrum.svg&oldid=179921766 [Accessed: 30 October 2019]

[10] Fricker HA. Lecture 7: Propagation, Dispersion and Scattering. Available from: https://topex.ucsd.edu/rs/Lec07.pdf [Accessed: 22 January 2020]

[11] Ito S, Murakami TN, Comte P, Liska P, Grätzel C, Nazeeruddin MK, et al. Fabrication of thin film dye sensitized solar cells with solar to electric power conversion efficiency over 10%. Thin Solid Films. 2008; **516**(14):4613-4619

[12] Office of Energy Efficiency and Renewable Energy. Solar radiation basics. Available from: https://www.energy.gov/eere/solar/articles/solar-radiation-basics [Accessed: 30 October 2019]

[13] GreenSpec. Solar hot water collector. Available from: http://www.greenspec.co.uk/building-design/solar-collectors/ [Accessed: 22 January 2020]

[14] Young WS, Abdullahi AA. Green electronics. Encyclopedia of Materials: Science and Technology. Reference Module in Materials Science and Materials Engineering. 2001:3645-3649. DOI: 10.1016/B978-0-12-803581-8.02029-4

[15] Perelaer J, Smith PJ, Mager D, Soltman D, Volkman SK, Subramanian V, et al. Printed electronics: The challenges involved in printing devices, interconnects, and contacts based on inorganic materials. Journal of Materials Chemistry. 2010;**20**:8446-8453

[16] Altay BN. Development and characterization of nano nickel-based conductive inks for flexographic printing of electronics and new interpretations of surface energies of solids [PhD dissertation]. Kalamazoo: Western Michigan University; 2018

[17] Willmann J, Stocker D, Dörsam E. Characteristics and evaluation criteria of substrate-based manufacturing. Is roll-to-roll the best solution for printed electronics? Organic Electronics. 2014; **15**(7):1631-1640

[18] Sagu JS, York N, Southee D, Wijayantha KGU. Printed electrodes for flexible, light-weight solid-state supercapacitors—A feasibility study. Circuit World. 2015;**41**(2):80-86

[19] Fang Z, Zhu H, Preston C, Han X, Li Y, Lee S, et al. Highly transparent and writable wood all-cellulose hybrid nanostructured paper. Journal of Materials Chemistry C. 2013;**1**:6191-6197

[20] Traille A, Kim S, Coustou A, Aubert H, Tentzeris MM. A conformal/rollable monolithic miniaturized ultra-portable ground penetrating radar using additive and inkjet printing. In: IEEE MTT-S International Microwave Symposium Digest. 2014. p. 6848398

[21] Chung O. Bendable, rollable, profitable. Taiwan Review. 2006;**56**(7):12-17

[22] Salleo A, Wong W. Flexible Electronics: Materials and Applications. Boston, MA, USA: Springer; 2009

[23] Kipphan H. Handbook of Print Media Technologies and Production Methods. Berlin, Germany: Springer; 2001

[24] Khan S, Lorenzelli L, Dahiya RS. Technologies for printing sensors and electronics over large flexible substrates: A review. IEEE Sensors Journal. 2015;**15**(6):3164-3185

[25] Gamota DR, Brazis P, Kalyanasundaram K, Zhang J, editors. Printed Organic and Molecular Electronics. Norwell, MA: Kluwer Academic Publishers; 2004

[26] Trudeau C, Bolduc M, Beaupré P, Topart P, Alain C, Cloutier S. Inkjet-printed flexible active multilayered structures. MRS Advances. 2017;**2**(18):1015-1020

[27] Trudeau C, Bolduc M, Beaupré P, Benavides-Guerrero J, Tremblay B, Cloutier S. Inkjet-printing of methylammonium lead trihalide perovskite as active layers for optoelectronic devices. MRS Advances. 2018;**3**(32):1837-1842

[28] Sekitani T, Nakajima H, Maeda H, Fukushima T, Aida T, Hata K, et al. Stretchable active-matrix organic light-emitting diode display using printable elastic conductors. Nature Materials. 2009;**8**(6):494-499

[29] Briseno AL, Mannsfeld SCB, Ling MM, Liu S, Tseng RJ, Reese C, et al. Patterning organic single-crystal transistor arrays. Nature. 2006;**444**(7121):913-917

[30] Krebs FC. Polymer solar cell modules prepared using roll-to-roll methods: Knife-over-edge coating, slot-die coating and screen printing. Solar Energy Materials and Solar Cells. 2009;**93**(4):465-475

[31] Comiskey B, Albert JD, Yoshizawa H, Jacobson J. An electrophoretic ink for all-printed reflective electronic displays. Nature. 1998;**394**(6690):253-255

[32] Altay BN, Husovska V, Pekarovicova A, Fleming PD. Formulating Pantone colors by unused base inks, formulation software and a spectrophotometer. Color Research & Applications. 2019;**44**:910-916

[33] Deore B, Paquet C, Kell AJ, Lacelle T, Liu X, Mozenson O, et al. Formulation of screen-printable Cu molecular ink for conductive/flexible/solderable Cu traces. ACS Applied Materials & Interfaces. 2019;**11**:38880-38894

[34] Huang X, Leng T, Zhang X, Chen JC, Chang KH, Geim AK, et al. Binder-free highly conductive graphene laminate for low cost printed radio frequency applications. Applied Physics Letter. 2015;**106**:203105

[35] 3D Printing Progress. Clariant acquires Nano-Silver Ink technology from Bayer. Available from: https://www.3dprintingprogress.com/articles/5345/clariant-acquires-nano-silver-ink-technology-from-bayer [Accessed: 22 January 2020]

[36] iDi Pac Ltd. The latest smart packaging technologies – some thoughts by Khoshal Mann. Available from: https://www.idipac.com/the-latest-smart-packaging-technologies-some-thoughts [Accessed: 22 January 2020]

[37] Available from: https://www.pinterest.ca/pin/87538786476697485/ [Accessed: 22 January 2020]

[38] Nth-Light. Printed LED light. Available from: https://e02fee51-b7de-42ea-b9b6-9315d13dec82.filesusr.com/ugd/99ec50_f4ead3ceea9e42bc88a07308fddbb960.pdf [Accessed: 22 January 2020]

[39] Kordás K, Mustonen T, Tóth G, Jantunen H, Lajunen M, Soldano C, et al. Inkjet printing of electrically conductive patterns of carbon nanotubes. Small. 2006;**2**(8–9):1021-1025

[40] Park M, Im J, Shin M, Min Y, Park J, Cho H, et al. Highly stretchable electric circuits from a composite material of silver nanoparticles and elastomeric fibres. Nature Nanotechnology. 2012;**7**(12):803-809

[41] Kattumenu R, Rebros M, Joyce M, Fleming PD, Neelgund G. Effect of substrate properties on conductive traces printed with silver-based flexographic ink. NPPRJ. 2009;**24**:101-106

[42] Lim S, Fleming PD, Joyce M. A study of the jetting evolution of nano copper ink and nano silver ink with inkjet. Journal of Imaging Science and Technology. 2013;**57**(2):1-8

[43] Lim S, Joyce M, Fleming PD, Aijazi AT, Atashbar M. Inkjet printing and sintering of nano copper ink. Journal of Imaging Science and Technology. 2013;**57**(5):1-7

[44] Määttänen A, Ihalainen P, Pulkkinen P, Wang S, Tenhu H, Peltonen J. Inkjet-printed gold electrodes on paper: Characterization and functionalization. ACS Applied Materials and Interfaces. 2012;**4**(2):955-964

[45] Husovska V, Pekarovic J, Pekarovicova A, Fleming PD. Capacitors out of recycled printed electronics paper substrates. Proceedings of the Technical Association of the Graphic Arts, TAGA. 2016:202-211

[46] Altay BN. Smart ink for flexo: Rossini scholarship winner examines printed electronics. Flexo. 2016;**41**(6):70-75

[47] Altay BN. Smart ink for flexo Rossini scholarship winner examines printed electronics. Flexo. 2016;**41**(5):26-32

[48] Altay BN, Jourdan J, Turkani VS, Dietsch H, Maddipatla D, Pekarovivoca A, et al. Impact of substrate and process on the electrical performance of screen-printed nickel electrodes: Fundamental mechanism of ink film roughness. ACS Applied Energy & Materials. 2018;**2018**(1):7164-7173

[49] Park S, Kim H. Flash light sintering of nickel nanoparticles for printed electronics. Journal of Thin Solid Films. 2014;**550**:575-581

[50] Alsaid DA, Rebrosova E, Joyce M, Rebros M, Atashbar M, Bazuin B. Gravure printing of ITO transparent electrodes for applications in flexible electronics. IEEE/OSA Journal of Display Technology. 2012;**8**(7):391-396

[51] Wang C, Tsai H, Hwang W. Direct printing of 1-D and 2-D electronically

conductive structures by molten lead-free solder. Materials. 2017;**10**(1):1-18

[52] Husovska V, Pekarovicova A, Fleming PD, Knox M, Fukushima H, Roberts K. Graphene inks for printed electronics. In: Edlund N, Lovrecek M, editors. Advances in Printing and Media Technology. Vol. 39. Darmstadt, Germany: International Association of Research Organizations for the Information, Media and Graphic Arts Industries; 2012

[53] Jeong SY, Kim SH, Han JT, Jeong HJ, Jeong SY, Lee G. Highly concentrated and conductive reduced graphene oxide nanosheets by monovalent cation-π interaction: Toward printed electronics. Advanced Functional Materials. 2012; **22**(15):3307-3314

[54] Hrehorova E, Rebros M, Pekarovicova A, Fleming PD, Bliznyuk VN. Characterization of conductive polymer inks based on PEDOT:PSS. TAGA Journal. 2008;**4**:219

[55] Ha YS, Park HG, Seo DS. Conductive polyaniline for potential application in anisotropic conductive films. Journal of Electronic Materials 2015;**44**(4):1200-1205

[56] Mäkelä T, Jussila S, Vilkman M, Kosonen H, Korhonen R. Roll-to roll method for producing polyaniline patterns on paper. In: Proceedings of the International Conference on Science and Technology of Synthetic Metals. 2003. pp. 41-42

[57] Xu Z, Ma Y. Development of iron-based superconducting thin films and coated conductors. Xiyou Jinshu/ Chinese Journal of Rare Metals. 2017; **41**(5):553-570

[58] Artamonovs O, Balodis G, Blums J. Electrical properties of the conductive pigments for flexible antennas. In: Proceedings of the Biennial Baltic Electronics Conference, BEC. 2014. pp. 145-148

[59] Pathak S, Pekarovicova A, Fleming PD. Zinc oxide based inks for semi conductive applications. Advances in Printing and Media Technology. 2015; **42**:195-204

[60] Everaerts K, Zeng L, Hennek JW, Camacho DI, Jariwala D, Bedzyk MJ, et al. Printed indium gallium zinc oxide transistors. Self-assembled nanodielectric effects on low-temperature combustion growth and carrier mobility. ACS Applied Materials and Interfaces. 2013;**5**(22):11884-11893

[61] Antoniadis H, Jiang F, Shan W, Liu Y. All screen printed mass produced silicon ink selective emitter solar cells. In: IEEE Photovoltaic Specialists Conference. 2010. pp. 1193-1196

[62] Ridley BA, Nivi B, Jacobson JM. All-inorganic field effect transistor fabricated by printing. Science. 1999; **286**(5440):746-749

[63] Sun Y, Kim HS, Menard E, Kim S, Adesida I, Rogers JA. Printed arrays of aligned GaAs wires for flexible transistors, diodes, and circuits on plastic substrates. Small. 2006;**2**(11): 1330-1334

[64] Trudeau C, Bolduc M, Beaupré P, Ka I, Asuo IM, Cloutier SG. Inkjet-printing of methylammonium lead trihalide perovskite-based flexible optoelectronic devices. In: 2018 International Flexible Electronics Technology Conference. 2018. pp. 1-6

[65] Haeberle T, Loghin F, Zschieschang U, Klauk H, Lugli P. Carbon nanotube thin-film transistors featuring transfer-printed metal electrodes and a thin, self-grown aluminum oxide gate dielectric. In: IEEE-NANO 2015—15th International Conference on Nanotechnology. 2015. pp. 160-163

[66] Wilhelm EJ, Jacobson JM. Direct printing of nanoparticles and spin-on-glasses by offset liquid embossing. Applied Physics Letters. 2004;**84**(18): 3507-3509

[67] Shin EY, Choi EY, Noh YY. Parylene based bilayer flexible gate dielectric layer for top-gated organic field-effect transistors. Organic Electronics: Physics, Materials, Applications. 2017;**46**:14-21

[68] Constantinescu C, Rapp L, Rotaru P, Delaporte P, Alloncle AP. Polyvinylphenol (PVP) microcapacitors printed by laser-induced forward transfer (LIFT): Multilayered pixel design and thermal analysis investigations. Journal of Physics D: Applied Physics. 2016;**49**(15):1-9

[69] Fan Y, Liu Y, Li H, Foulds IG. Printed wax masks for 254 nm deep-UV pattering of PMMA-based microfluidics. Journal of Micromechanics and Microengineering. 2012;**22**(2)

[70] Moon HC, Lodge TP, Frisbie CD. Solution-processable electrochemiluminescent ion gels for flexible, low-voltage, emissive displays on plastic. Journal of the American Chemical Society. 2014;**136**(9): 3705-3712

[71] Hwang HJ, Hsu SW, Wang CS. Low dielectric thermoset from redistributed poly(phenylene oxide). Journal of Macromolecular Science, Part A: Pure and Applied Chemistry. 2008;**45**(12): 1049-1056

[72] Allen M, Lee C, Ahn B, Kololuoma T, Shin K, Ko S. R2R gravure and inkjet printer RF resonant tag. Microelectronic Engineering Journal. 2011;**88**:3293-3299

[73] Kojima A, Teshima K, Shirai Y, Miyasaka T. Organometal halide perovskites as visible-light sensitizers for photovoltaic cells. Journal of American Chemical Society. 2009; **131**(17):6050-6051

[74] Saliba M, Matsui T, Seo J, Domanski K, Correa-Baena J, Nazeeruddin M, et al. Cesium-containing triple cation perovskite solar cells: Improved stability, reproducibility and high efficiency. Energy & Environmental Science. 2016;**9**(6): 1989-1997

[75] Lee Y, Kwon J, Hwang E, Ra CH, Yoo WJ, Ahn JH, et al. High performance perovskite-graphene hybrid photodetector. Advanced Materials. 2015;**27**(1):41-46

[76] Brenner TM, Egger DA, Rappe AM, Kronik L, Hodes G, Cahen D. Are mobilities in hybrid organic-inorganic halide perovskites actually "high"? Journal of Physical Chemistry Letters. 2015;**6**(23):4754-4757

[77] MacDonald W. Engineered films for display technologies. Journal of Materials Chemistry. 2014;**14**:4-10

[78] Siegel AC, Phillips ST, Dickey MD, Lu N, Suo Z, Whitesides GM. Foldable printed circuit boards on paper substrates. Advanced Functional Materials. 2010;**20**:28-35

[79] Garner S, Glaesemann S, Li X. Ultra-slim flexible glass for roll-to-roll electronic device fabrication. Applied Physics A. 2014;**116**:403-407

[80] Stoppa M, Chiolerio A. Wearable electronics and smart textiles: A critical review. Sensors (Switzerland). 2014; **14**(7):11957-11992

[81] Hong Y, Chung S, Kattamis A, Cheng I, Wagner S. Technical issues of stainless steel foil substrates for OLED display applications. Proceedings of SPIE. 2007;**6655**:66550N-665501N

[82] Ji SY, Chang WS, Choi WH, Kim HW, Cho S, Jeon J. Fully solution processable fabrication of multi-layered circuits on a flexible

substrate using laser processing. Materials. 2018;**11**(2):1-9

[83] Akhavan V, Schroder K, Farnsworth S. Photonic curing of inkjet printed films. In: Handbook of Industrial Inkjet Printing: A Full System Approach. Weinheim, Germany: Wiley-VCH Verlag GmbH & Co. KGaA; 2018

[84] Zhang LC, Sun X, Hu Z, Yuan CC, Chen CH. Rice paper as a separator membrane in lithium-ion batteries. Journal of Power Sources. 2012;**204**: 149-154

[85] Li Y, Chen Q, Li L, Wang S, Wang Z. Review of damage problems of the soft substrate interlayer film. Vibroengineering Procedia. 2015;**5**: 297-302

[86] Suganuma K. Introduction to Printed Electronics. New York, USA: Springer Science+Business Media; 2014

[87] Cronin HM, Stoeva Z, Brown M, Shkunov M, Silva SRP. Photonic curing of low-cost aqueous silver flake inks for printed conductors with increased yield. ACS Applied Materials & Interfaces. 2018;**10**:21398-21410

[88] Zhang Y, Zhang J. Sintering phenomena and mechanical strength of nickel based materials in direct metal laser sintering process—A molecular dynamics study. Journal of Materials Research. 2016;**31**(15):2233-2243

[89] Niittynena J, Abbelb R, Mäntysalo M, Perelaer J, Schubert U, Lupo D. Alternative sintering methods compared to conventional thermal sintering for inkjet printed silver nanoparticle ink. Thin Solid Films. 2014; **556**:452-459

[90] Schroder K. Mechanisms of photonic Curing™: Processing high temperature films on low temperature substrates. In: Technical Proceedings of the NSTI Nanotechnology Conference and Expo. NSTI-Nanotech; 2011. pp. 220-223

[91] Al-Fuqaha A, Guizani M, Mohammadi M, Aledhari M, Ayyash M. Internet of Things: A survey on enabling technologies, protocols and applications. IEEE Communications Surveys & Tutorials. 2015;**17**(4):2347-2376, Fourth quarter

[92] Sowade E, Polomoshnov M, Baumann RR. The design challenge in printing devices and circuits: Influence of the orientation of print patterns in inkjet-printed electronics. Organic Electronics. 2016;**37**:428-438

[93] Marques GC, von Seggern F, Dehm S, Breitung B, Hahn H, Dasgupta D, et al. Influence of humidity on the performance of composite polymer electrolyte-gated field-effect transistors and circuits. IEEE Transactions on Electron Devices. 2019; **66**(5):2202-2207. DOI: 10.1109/ TED.2019.2903456

[94] Trudeau C. Printed electronics prototyping of flexible optoelectronic methylammonium trihalide perovskite-based devices [PhD dissertation]. Montreal: École de technologie supérieure; submitted 2020

[95] Brewer D. AI needs printed electronics for sensor arrays. Internet of Things Designline. EE Times. Available from: https://www.eetimes.com/ai-needs-printed-electronics-for-sensor-arrays/ [Accessed: 22 January 2020]

Chapter 2

Techno-Economic Optimization and Benchmarking of a Solar-Only Powered Combined Cycle with High-Temperature TES Upstream the Gas Turbine

Fritz Zaversky, Iñigo Les, Marcelino Sánchez, Benoît Valentin, Jean-Florian Brau, Frédéric Siros, Jonathon McGuire and Flavien Berard

Abstract

This work presents a techno-economic parametric study of an innovative central receiver solar thermal power plant layout that applies the combined cycle (CC) as thermodynamic power cycle and a multi-tower solar field configuration together with open volumetric air receivers (OVARs). The topping gas turbine (GT) is powered by an air–air heat exchanger (two heat exchanger trains in the case of reheat). In order to provide dispatchability, a high-temperature thermocline TES system is placed upstream the gas turbine. The aim is threefold, (i) investigating whether the multi-tower concept has a techno-economic advantage with respect to conventional single-tower central receiver plants, (ii) indicating the techno-economic optimum power plant configuration, and (iii) benchmarking the techno-economic optimum of the CC plant against that of a conventional single-cycle Rankine steam plant with the same receiver and TES technology. It is concluded that the multi-tower configuration has a techno-economic advantage with respect to the conventional single-tower arrangement above a total nominal solar power level of about 150 MW. However, the benchmarking of the CC against a Rankine single-cycle power plant layout shows that the CC configuration has despite its higher solar-to-electric conversion efficiency a higher LCOE. The gain in electricity yield is not enough to outweigh the higher investment costs of the more complex CC plant layout.

Keywords: concentrated solar power, solar combined cycle, multi-tower central receiver, open volumetric air receiver (OVAR)

1. Introduction

Solar thermal power, also known as concentrated solar power (CSP) or solar thermal electricity (STE), can be considered as a highly promising technology when it comes to dispatchable and thus grid-friendly supply of renewable electricity. This is due to the possibility of thermal energy storage (TES), the key advantage over other renewable technologies (such as wind or photovoltaic), which enables the decoupling between solar energy collection and electricity production. Given the abundant amount of solar power available for terrestrial solar collectors (85 PW) [1], which exceeds the current world's power demand (15 TW) several thousand times [1], CSP is a highly promising and flexible alternative to conventional fossil-fuel technologies, setting new standards in terms of environmental impact, sustainability, safety, and thus quality of life.

Until recently, the cost of electricity generation for CSP (≈ 14 c€/kWh [2]) has been clearly above conventional technology and other renewables (wind and photovoltaic reach 6 c€/kWh on average [2]). Nevertheless, considering latest bids, cost targets as low as 7 c$/kWh [3] seem to be realistic for mature CSP technology. However, rather than comparing the pure cost, one should compare the true value of CSP for grid operation and capacity [2] when considering an increasing fraction of not-dispatchable renewables. It is important to note that CSP should not be seen as a competing technology with photovoltaic or wind power. It should rather be seen as an enabling technology of not-dispatchable renewable power generation, as a key feature of CSP is its cheap thermal energy storage, guaranteeing dispatchability.

This work focuses on central receiver CSP plants [4] analyzing the economic competitiveness of the combined cycle (topping GT plus bottoming steam Rankine) with respect to conventional single-cycle Rankine steam technology. The combined cycle technology [5–7] is well known from conventional fossil-fired power generation, reaching cycle efficiencies exceeding 60% on a lower heating value basis [8]. However, these best of class efficiencies are obtained with latest fossil-fired gas turbine (GT) technology, achieving turbine inlet temperatures (TITs) of up to ≈1500°C. It is clear that such high TITs can only be achieved with (i) internal combustion and (ii) turbine blade cooling and single-crystal superalloy turbine blades, furthermore coated with low conductivity ceramics. For the application of solar combined cycles, the TIT has to be significantly lower for two reasons: (i) the optimum receiver operating temperature (≈ TIT) that optimizes solar-to-electric conversion efficiency is a function of the receiver's thermal efficiency and tends to be at around 1000°C [9, 10] depending on the receiver technology; (ii) for solar combined cycles, the concept of externally heated gas turbines [11] has to be exploited, which limits the maximum achievable TIT to lower values (≈ 900–1000°C) in any case. Additionally, cheaper designs for the turbine should be used in order to keep costs down, i.e., uncooled turbine blades, which should be achievable with expected optimum receiver working temperatures of ≈ 1000°C [9]. Furthermore, advanced gas turbine architectures [12] such as reheat will be necessary to achieve good GT efficiencies, despite low TITs. Reheated gas turbines have already been treated in previous works. The main motivations are (i) to keep the average temperature of heat supply high and (ii) to introduce an additional flexibility regarding turbine exit temperature (TET) (the heat recovery steam generator inlet temperature), despite high compressor pressure ratios [13]. In particular, the expansion ratio of the second turbine stage can be specifically designed, so that the resulting TET optimizes the overall combined cycle performance. Note that the higher the heat recovery steam generator's inlet temperature is, the higher are the conversion efficiency and power output of the bottoming cycle and vice versa.

It is clear that the solar receiver unit is the key component of a solar-powered combined cycle plant, since it is of upmost importance to achieve very good solar receiver efficiencies at highest operating temperatures (≈ 1000°C). So far, pressurized air receivers have been the design principle for solar-powered gas turbines, since the solar receiver has to provide heat to a pressurized air stream coming from the gas turbine's compressor. Several previous research projects have already endeavored to design such a demanding component, which has to operate under very high solar flux (≈ 0.5–1 MW/m^2), at high temperatures (> 900°C), and in addition at pressures over 6 bar. The first developments started in the 1980's with metallic and ceramic tubular designs [14]. This approach showed however durability issues and also low efficiencies because of the low heat transfer coefficient of air. Therefore, pressurized volumetric receivers [15] appeared to be a promising alternative as they increased the heat transfer area. However, durability issues and size limitations of the needed quartz glass window have hindered their commercial application so far. For this reason, the idea of pressurized tubular or opaque-heat-exchanger-type receivers was revisited by several research groups. For example, Grange et al. [16] investigated a modular metallic absorber located at the back of a cavity. The maximum air outlet temperature was reported to be 750°C. Korzynietz et al. [17] developed a pre-commercial scale metallic tubular cavity receiver achieving thermal efficiencies between 71.3 and 78.1% at the maximum air outlet temperature of 800°C.

However, although dispatchability is the key advantage of CSP (due to cost-effective thermal energy storage) and its best argument to justify higher costs than PV or wind energy, only a few works have covered the integration of high-temperature TES upstream the solar combined cycle. Since only pressurized receivers have been applied so far in the context of solar-powered combined cycles, previous works have proposed the application of pressurized regenerative TES systems [16, 18], which have clear limitations regarding cost-effective large-scale deployment.

The aim of this work is therefore to present an innovative plant layout (as shown in **Figure 1**) that not only avoids the design challenges related to pressurized receivers but also allows the integration of an atmospheric air-based high-temperature TES system upstream the combined cycle. In particular, this work proposes the application of the open volumetric air receiver technology [15], which has already been demonstrated successfully at pre-commercial scale [19], in

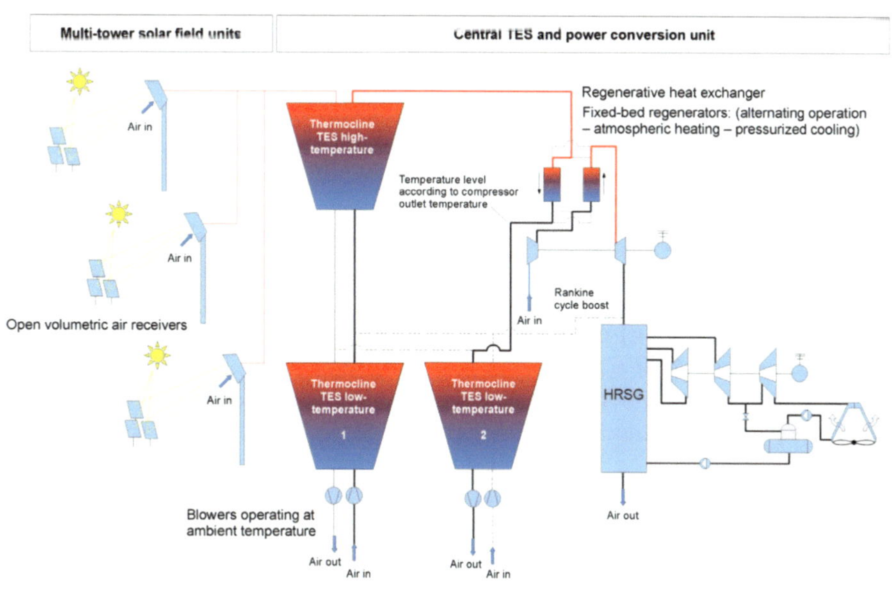

Figure 1.
Solar-powered combined cycle scheme with open volumetric air receiver and high-temperature TES (without reheat in the Brayton cycle). The low-temperature TES enables regenerative use of return air heat.

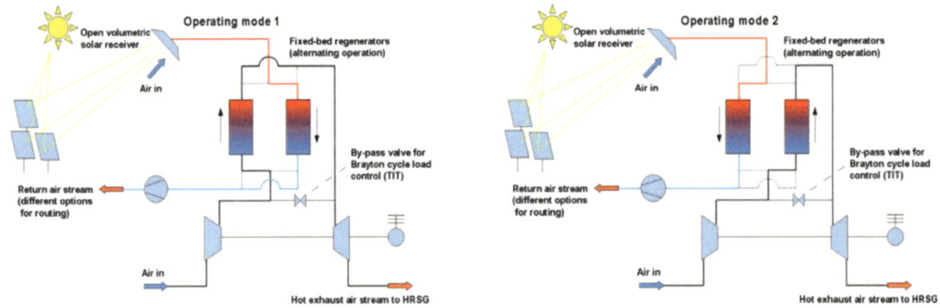

Figure 2.
Innovative coupling of open volumetric air receiver and Brayton cycle.

combination with a regenerative system working in alternating modes (atmospheric heating, pressurized cooling; see **Figure 2**) in order to power a solar-only combined cycle. This approach decouples the high-temperature and the high heat flux part (solar receiver) from the high pressure part (compressed air stream of the Brayton cycle) via an air–air regenerative heat exchanger. Thus, upstream the combined cycle, the well-proven and relatively cheap regenerator-type heat storage known from the so-called Cowper Stoves [7, 20, 21] can be used. The regenerative matrix consists of refractory bricks with channels in between where the gas is flowing, transferring heat to the bricks or vice versa [20]. This type of regenerative heat storage has already been demonstrated successfully at pilot plants [22–24] for the application of CSP. The big advantages of this technology are (i) a simple design with very low technological risk and (ii) low costs ($\approx$17 €/kWh$_{th}$ [25]).

The temperature level of the "cold" return air stream leaving the air–air heat exchange system is a function of compressor outlet temperature (i.e., compressor pressure ratio and ambient temperature) and the exit temperature of the first turbine stage, in the case of reheat. The resulting air-return temperature level is too high for efficient blower operation and the recirculation to the high-temperature TES or the receiver is thus not feasible. A low-temperature air/rock thermocline TES is thus proposed in order to reuse the return air heat in regenerative manner. In order to keep air transport parasitic power consumption acceptable, the operating temperature of the blower should be kept at ambient temperature level.

A related work [10] has shown that the thermodynamic performance of this power plant layout, having an effective mean solar flux concentration ratio of 500, optimizes at a receiver outlet temperature ($\approx$TIT) of about 1050°C and a Brayton cycle pressure ratio of 14 (reheat ratio K=0.75), resulting in about 29.6% peak solar-to-electric conversion efficiency. The present work continues the research, focusing on the techno-economic optimization and benchmarking of the proposed power plant layout. The aim is threefold, (i) investigating whether the multi-tower concept has a techno-economic advantage with respect to conventional single-tower central receiver plants, (ii) indicating the techno-economic optimum of the power plant size (the number of towers, solar field size, solar multiple, and hours of TES), and (iii) benchmarking the techno-economic optimum of the solar-powered combined cycle plant against that of a conventional single-cycle Rankine steam plant [22] with the same receiver (but lower operating temperature, $\approx$ 800°C) and TES technology (see **Figure 3**).

1.1 Advantages and limitations of air as heat transfer fluid (HTF)

Probably the most important advantage of air as HTF is that it has no temperature limit. Thus, overheating and freezing are no issue, in contrast to the application

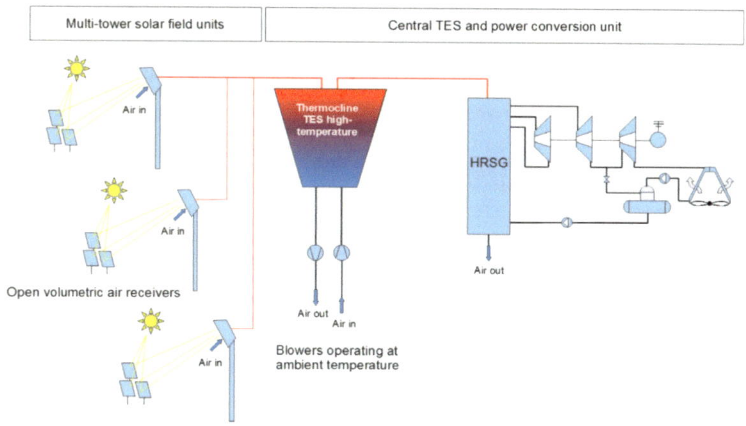

Figure 3.
Single cycle Rankine scheme—Optimum receiver outlet temperature is 800°C.

of molten salts or thermal oil. For this reason, no expensive heat-tracing equipment is needed, and furthermore the application of air as HTF allows the implementation of advanced power cycles such as the combined cycle (topping gas turbine and bottoming Rankine cycle), which requires temperatures of cycle heat input in the range between 800 and 1100°C (the optimum gas turbine inlet temperature depends on the receiver efficiency).

The second advantage is that air is freely available and obviously nontoxic to human kind and nature. Thus, no investment or cost of replacement during power plant life time must be considered. However, it may be necessary to thoroughly clean (filter) the air after recirculation in the heat transport circuit before its release to the ambient, as particulate matter coming from piping material, especially from high-temperature insulation fibers, must not be released to the atmosphere. If released it would represent an important health hazard.

The third advantage is that very cost-effective thermal energy storage technology is available when using atmospheric air as HTF. Here, the well-proven and relatively cheap regenerator-type heat storage known from the so-called Cowper stoves [20, 21], which are applied together with blast furnaces, can be used. These high-temperature regenerators work at temperatures of up to 1250°C [20] providing hot gas at constant temperature to the furnace. The outlet temperature of the regenerators is typically controlled by adding a variable flow of cold air at the outlet [21]. The regenerative matrix consists of refractory bricks with channels in between where the gas is flowing, transferring heat to the bricks or vice versa [20]. This type of regenerative heat storage has already been demonstrated successfully at pilot plants [22, 23] for the application of CSP. The big advantages of this technology are (i) a simple design with very low technological risk and (ii) low costs [25]. Possible storage vessel core geometries are packed beds, consisting of spheres or broken particles, stacks of plates, perforated bricks, or extruded shapes [26]. In packed beds, the efficiency of thermal energy storage depends on the heat transfer between the air and the filler material, as well as on the reached stratification or thermocline. Good heat transfer and limited heat transport within the solid storage media that enhances thermal stratification is reached by porous structures [27]. A truncated conical shape of the container, coupled with subterranean location, may alleviate problems such as rock fracture and tank deformation. The inclined walls reduce the mechanical constraints by guiding the rocks upwards during thermal expansion [24].

As indicated above, air as HTF has important advantages, which however are limited by the fact that air is not very suitable when it comes to heat transport over longer distances and heat transfer, especially at atmospheric pressure. The heat transport issue of atmospheric air circuits is caused by the very low density of air at high temperatures (≈ 0.28 kg/m^3 at 1000°C), which leads to high flow velocity and thus elevated pressure drops (high blower power consumption). Therefore, the air piping system must provide sufficiently large flow cross sections in order to keep flow velocity in the order of magnitude of 15–30 m/s. This requirement leads to large diameter piping that needs to be insulated internally as high-temperature alloys are too expensive.

However, not only the associated pressure drop (parasitic power consumption) is an issue when applying air as HTF. Also thermal losses and thermal inertia effects are important design aspects for air-based CSP plants.

All the abovementioned disadvantages have to be properly addressed for the scale-up to commercial size of atmospheric air-based CSP plants. As described later on, the upscaling of CSP plants is important, as specific power cycle costs (USD/kWe) are significantly lower when moving to higher nominal power output. Therefore, CSP plants have increased in size recently. When upscaling a CSP plant to several hundreds of MWe, several heliostat fields and receivers may be needed to provide heat to the same power cycle: in such a case, the applied HTF needs to transport heat over very long distances (1–2 km). However, when applying air as HTF, thermal losses as well as thermal inertia effects (because of cooldown at night) are a major hurdle to its large-scale commercial implementation. In the following, the performance of an 800 m high-temperature air duct will be given, transporting the thermal power delivered by a 51 MWth heliostat field (Case C, **Table 1**).

Assuming a receiver outlet temperature of 1000°C and a corresponding receiver efficiency of 0.754 results in a nominal air mass flow of about 36 kg/s. For the design of the needed piping system, several techno-economic considerations need to be taken into account. The piping inner diameter defines the flow cross section and the outer circumferential surface area which needs to be thermally insulated. The ratio of flow cross area to the outer surface area (that defines piping and insulation material mass) becomes higher for larger inner diameters. Therefore, the inner diameter should be chosen as big as possible, i.e., considering manufacturing and structural strength limitations. In principle, circular duct geometry is the better choice as outer surface area per square meter of flow cross section is lower than in the case of rectangular cross section.

For the specific example, an inner piping diameter of 3 m has been chosen. The default thickness and thermal conductivity of the thermal insulation are 0.6 m and 0.035 W/(m K), respectively. The heat transfer coefficient between ambient air and outer pipe surface is assumed to be ≈ 10 W/(m^2 K). For the default case (Case 1 in **Table 2**), the given settings result in a flow velocity of 19 m/s and a total pressure drop [28] of about 0.22 kPa (800 m total piping length). The thermal loss to the ambient air (at 25°C) causes a temperature reduction of about 12°C, down to 988°C. As **Table 2** further indicates, when increasing the mass flow rate, the temperature drop reduces; however pressure drop more than doubles. Thus, one measure to reduce the temperature reduction is to increase mass flow rate, however, with the cost of higher parasitic power consumption. Obviously, since the thermal losses remain practically constant (the same temperature difference to ambient and overall heat transfer coefficient remains almost constant), a higher mass flow rate translates to lower temperature difference in the air flow. The second measure to reduce temperature drop is to increase insulation thickness (see **Table 3**), which however is very expensive.

	Solar field power classes				
	A	**B**	**C**	**D**	**E**
Nominal solar power (MW)	5.666 MW (North)	17 MW (North)	51 MW (North)	153 MW (Surround)	459 MW (Surround)
Mirror area (m^2)	7248	21,960	68,560	221,640	710,240
Annual optical efficiency (−)	0.7782	0.7644	0.7368	0.6827	0.6486
Field diameter (m)	240	380	750	1300	2400
Tower height (m)	50	80	100	150	200
Nominal solar power (MW)	**Tower configurations analyzed and corresponding LCOE (c$/kWh (CC)) (c$/kWh (RC))**				
17	**3 A** (CC = 19) (RC = 20.4)	**1 B** (CC = 16.7) (RC = 17.4)	—	—	—
34	**6 A** (CC = 17.7) (RC = 18.2)	**2 B** (CC = 15.6) (RC = 15.7)	—	—	—
51	**9 A** (CC = 17) (RC = 16.9)	**3 B** (CC = 15.1) (RC = 14.7)	**1 C** (CC = 13.8) (RC = 13.1)	—	—
102	—	**6 B** (CC = 14.1) (RC = 13.6)	**2 C** (CC = 13.6) (RC = 13)	—	—
153	—	**9 B** (CC = 13.6) (RC = 12.5)	**3 C** (CC = 13.4) (RC = 12.5)	**1 D** (CC = 13.8) (RC = 12.6)	—
204	—	—	**4 C** (CC = 13.2) (RC = 11.8)	—	—
306	—	—	**6 C** (CC = 12.6) (RC = 11.5)	**2 D** (CC = 14.6) (RC = 13.7)	—
459	—	—	**9 C** (CC = 12.9) (RC = 12.3)	**3 D** (CC = 14.4) (RC = 14)	**1 E** (CC = 14.7) (RC = 14.2)
612	—	—	—	**4 D** (CC = 14.5) (RC = 14.1)	—

Table 1.
Solar field base modules and multi-/single-tower configurations analyzed.

This analysis shows that a temperature drop of ≈10°C is a reasonable assumption for the multi-tower concepts analyzed later on.

Last but not the least, the second issue, as mentioned above, is related to low heat transfer performance when applying air as HTF. For this reason, heat exchangers need to have a very large area of heat transfer in order to counterbalance this drawback. This is the reason why heat recovery steam generators of combined cycle plants are very bulky and represent an important share of the power block's CAPEX. The same holds for the air–air heat exchanger that is needed for the CAPTure power plant layout (see **Figure 1**). Here, the motivation is to reduce cost

	Case 1	Case 2	Case 3	Case 4	Case 5	Case 6
Roughness (m)	0.001	0.001	0.001	0.001	0.001	0.001
Inner diameter (m)	3	3	3	3	3	3
Insulation thickness (m)	0.6	0.6	0.6	0.6	0.6	0.6
Air flow (kg/s)	36	39	42.4	46	49.5	56.6
Inlet temperature (°C)	1000	1000	1000	1000	1000	1000
Outlet temperature (°C)	988	989	990	991	992	992.5
Pressure drop (kPa)	0.22	0.25	0.3	0.35	0.4	0.52
Velocity (m/s)	19	20	22	24	26	29
Inner wall heat transfer coefficient (W/m^{2}K)	12.5	13	14	15	16	18

Table 2.
Variation of mass flow—The same piping diameter and insulation.

	Case 1	Case 2a	Case 3a	Case 4a	Case 5a
Roughness (m)	0.001	0.001	0.001	0.001	0.001
Inner diameter (m)	3	3	3	3	3
Insulation thickness (m)	0.6	0.8	1	1.2	1.4
Air flow (kg/s)	36	36	36	36	36
Pressure drop (kPa)	0.22	0.22	0.22	0.22	0.22
Inlet temperature (°C)	1000	1000	1000	1000	1000
Outlet temperature (°C)	988	990.5	992	992.7	993

Table 3.
Variation of insulation thickness—The same mass flow and diameter.

and heat exchanger size by applying a regenerative heat exchange system (atmospheric heating, pressurized cooling).

1.2 The high-temperature thermocline thermal energy storage system upstream the gas turbine

The challenges of employing air as HTF are not only related to the HTF transport itself but also to the solar receiver design. Advanced power cycles are typically highly recuperative, since the average temperature of heat input to the cycle must be maintained high. This necessity implies that the HTF temperature interval in which heat is supplied to the cycle is typically small. In the case of the combined cycle, the temperature difference of the HTF is determined by the turbine inlet temperature and the compressor exit temperature (CET), which is a function of Brayton cycle pressure ratio. In the case of reheat, the HTF temperature difference is also determined by the reheat pressure level, i.e., the turbine exit temperature (TET) of the first turbine stage. Ideally, these two temperature levels (CET and TET) should be similar, in order to reduce losses when mixing the two streams (effective HTF return temperature). The effective HTF return temperature is the temperature at which the HTF leaves the power block after all parallel mass flow streams are merged (mixing temperature). At this temperature level, the HTF would then be recirculated to the solar receiver and again heated to the nominal receiver outlet temperature, if an efficient and cost-effective receiver design was available for the case of atmospheric air.

However, in contrast to tubular receivers as used with molten salts, for example, the HTF recirculation, for the case of atmospheric air and open volumetric receivers, is not trivial. Previous research projects [29, 30] have targeted the design of open volumetric air receivers with air recirculation, however with limited success since only about 50% of the return air could be successfully recirculated [29]. Therefore, from a thermodynamic point of view, it would be more efficient to store the low-temperature heat of the return air for later use. Also, considering advanced power cycles with elevated HTF return temperatures (see the above), it would be too detrimental for the global performance, if relatively hot air was blown back to the receiver(s), especially when considering a multi-tower concept. It is clear that the operating temperature of air blowers must be as close as possible to ambient temperature, in order to keep a reasonable density.

Therefore, the present work considers the use of the return air heat in regenerative manner, applying a low-temperature air/rock thermocline TES (see **Figure 1**). The basic TES subunit consists of three packed beds, one high-temperature and two low-temperature units. Two low-temperature TES units are required because a packed bed cannot be charged and discharged at the same time. Hence, while discharging the TES system, the first low-temperature unit preheats ambient air until the nominal power cycle HTF return temperature is reached. Then, the preheated air enters the high-temperature storage unit where the air is heated to the nominal power cycle inlet temperature. Next, the HTF enters the power cycle, delivers part of its heat to the cycle, and finally charges the second low-temperature storage unit before being rejected to the atmosphere at a temperature very close to ambient conditions. During TES charging, only the high-temperature thermocline unit is being charged, that is, once the nominal receiver outlet temperature is approached at the bottom part of the high-temperature (HT) bed (cutoff condition), the TES system is fully charged. This also means that the low-temperature (LT) TES units need to be designed with larger bed heights (at same diameter and filling material characteristics) so that the blowers always operate close to ambient temperature and no heat is lost to the ambient. When upscaling the thermal storage capacity, several of such three bed subunits (1 HT TES + 2 LT TES) need to be applied. Additionally, in order to keep the system balanced, so that always one LT TES system is empty before the storage system is being discharged, one of the LT TES units needs to be discharged by boosting/powering the bottoming Rankine cycle. Note that the (dispatchable) boost operation is preferred over the direct boost during diurnal charging as the air mass flow is constant and does not depend on current solar irradiance (receiver temperature control). Therefore, the Rankine cycle can be run at constant load. Also, in order to keep the thermocline in the LT TES system balanced, its operation needs to be switched from time to time, so that not always the same LT TES unit is being charged by the HTF return stream. Having separate HT and LT TES units allows the application of temperature-specific materials for the packed bed as well as for the internal insulation of the tank, which allows to design the LT units much more economically.

The design of the packed beds would be equivalent to the system proposed in Refs. [23, 24].

Last but not the least, one remark needs to be given on how to operate the plant during the day. On the one hand, there is the possibility to only operate the power block during the night or only at times of high electricity prices, powering it via the TES only. In this case, direct solar operation is not implemented. On the other hand, if direct solar operation during the day is wanted in order to increase the capacity factor of the power block, two possibilities exist: (i) only the Rankine cycle is operated during the day, which eliminates the need for a regenerative LT TES during the day, and also allows a reduced receiver operating temperature during the day, once the HT TES is full; (ii) the combined cycle is operated during the day, channeling the HTF return stream to the bottoming Rankine cycle, which significantly increases Rankine cycle output (i.e., its nominal power) and leads to reduced overall conversion efficiency, since it is not a true combined cycle. The operating principle of choice will depend on the specific electricity market.

1.3 The motivation for a reheated Brayton cycle and its application in the context of combined cycle power generation

As it is well known according to the principles of thermodynamics, reheat increases the average temperature of heat supply and thus increases the conversion efficiency from heat to mechanical work (Carnot). However, since a second heat input adds additional pressure losses, higher compressor pressure ratios are required to offset the performance penalty. This requirement is even more relevant for a solar-powered thus an externally heated Brayton cycle. Reheated gas turbines have been already treated in previous works. The main motivations are (i) to keep the average temperature of heat supply high and (ii) to introduce an additional flexibility regarding turbine exit temperature (the heat recovery steam generator inlet temperature), despite high compressor pressure ratios [13, 31]. In particular, the expansion ratio of the second turbine stage can be specifically designed, so that the resulting TET optimizes the overall combined cycle performance. Note that the higher the heat recovery steam generator's inlet temperature is, the higher is the conversion efficiency of the bottoming cycle and vice versa. As proposed by Siros and Fernández-Campos [12], the reheat pressure level will be defined by a dimensionless parameter K (reheat ratio), which determines the ratio of pressure ratios of both turbine stages:

$$K = \frac{pressure\ ratio\ of\ first\ stage}{pressure\ ratio\ of\ second\ stage} = \frac{\frac{p_{t1\ i}}{p_{t1\ o}}}{\frac{p_{t2\ i}}{p_{t2\ o}}} = \frac{p_{t1\ i}}{p_{t1\ o}} \cdot \frac{p_{t2\ o}}{p_{t2\ i}} \qquad (1)$$

Here, three considerations must be kept in mind:

i. The reheat ratio K is a key parameter concerning Brayton cycle performance (on its own) as well as combined cycle performance; nevertheless, it has different optimums for the single cycle and the combined cycle. The lower the pressure ratio of the second turbine stage is, the higher the turbine exit temperature, i.e., HRSG inlet temperature, and thus the higher the efficiency of the bottoming Rankine cycle, but the lower the Brayton cycle performance. As shown in Ref. [10], solar combined cycle performance optimizes in the interval of $0.5 < K < 1.25$. The optimum value of K depends on concentration ratio, the corresponding optimum TIT, and HRSG efficiency. Note that Brayton single-cycle performance optimizes for values of K lower than in the case of combined cycle (see Ref. [32]).

ii. Furthermore, the lower the pressure ratio of the second turbine stage is, the lower is the reheat pressure level and thus the bulkier and more expensive the second HTF-to-working-fluid heat exchanger will be. And the pressure drop would increase. Thus, there is clearly a lower practical limit for the second turbine stage's pressure ratio.

iii. Having higher pressure ratios in the first turbine stage means lower TET at the first stage and thus corresponds to a lower return temperature of the TES medium (thus higher ΔT for the TES). Thus, higher pressure ratios in the first turbine stage are not only preferred in terms of Rankine cycle performance (see point (i)) but also regarding integration with thermal energy storage (second heat exchanger train).

In summary, a reheated Brayton cycle is absolutely interesting for the application of CSP as it allows fair conversion efficiencies despite low TITs. Reheat may increase solar-to-electric performance by up to about 2.7 percentage points [10]. Another argument for reheat is that the optimum cycle pressure ratio is much higher than in the case of a simple Brayton cycle. This means that the compressor exit temperature is higher, which allows for a higher effective HTF return temperature (low-temperature TES inlet temperature) and thus a more efficient boost of the bottoming Rankine cycle (see **Figure 1** dashed lines indicating periodical Rankine cycle boost in order to discharge the LT TES).

1.4 Motivations for multi-tower CSP plant arrangements

Basically, there are two options for multi-tower arrangements, (i) a simple multiple placement of identical solar field tower-receiver units, where each heliostat field concentrates solar radiation onto its corresponding receiver (tower) only or (ii) a special arrangement of multiple towers and heliostat fields, where heliostats of one field may point on different receivers (towers) as function of current solar position in order to optimize the overall optical efficiency. We propose to refer to the latter option "multi-tower multi-aiming" configuration, and to the first option "multi-tower assigned-aiming" configuration. The "multi-tower assigned-aiming" configuration is considered in this work.

Multi-tower arrangements with compact heliostat fields have significant advantages regarding solar field efficiency, atmospheric attenuation, and solar flux control at the receivers. This is because heliostats placed closer to the tower have higher optical efficiencies than those placed at the peripheral areas of the field [33]. Longer slant ranges, which may reach 1.5 km and more at large-capacity single-tower concepts [34, 35], already cause losses of up to 10% [36] purely considering atmospheric attenuation losses, not to mention spillage losses and increasingly challenging solar flux control. Thus, when upscaling a central receiver plant, there is a point where the heliostat field becomes simply too large. For this reason, a multi-tower approach is a promising way when going for very high capacity power tower plants, where Rankine cycle power blocks become more efficient and also cheaper per installed MW.

Additionally, the ongoing transformation from centralized conventional power generation to decentralized power supply with a high share of renewables calls for smaller modular units that can be easily adapted to the specific power demand. The general trend is expected to go towards more but lower capacity power generation units with lower capital risk and lower amount of initial investment [37]. However, so far in the case of CSP, current cost reduction trends are mainly driven by the increase of the nominal size of the main components (especially the power block) in order to reduce the cost of electricity production, as specific power cycle costs ($/kWe) significantly reduce for large power ratings. This trend would change if we paid attention to the needs of the consumers and the changing, more and more decentralized electricity grid. Also, a modular design may reduce the perceived risk of the technology, giving the technology a better access to financing.

The principal problem in this context is that there is no power cycle available so far that is also cost-effective and efficient in smaller power classes. Typically, specific costs ($/kWe) of gas turbines and Rankine cycles increase significantly for small power classes, and conversion efficiencies also decrease. If there was a cheap and efficient power cycle available in the power class below or around 10 MWe, solar power towers would be very compact plants, as the optical efficiency and consequently the solar-to-thermal efficiency are best for small solar fields. The only way in order to combine (i) good solar-to-thermal efficiency and (ii) an efficient power cycle (i.e., to maximize solar-to-electric energy conversion) is the application of multi-tower power plant concepts.

2. Cost review of power plant components

In order to perform a serious techno-economic study, the fundamental step is to collect realistic cost estimates for all power plant components. Therefore, a detailed literature search has been conducted collecting available cost data and also comparing them in order to guarantee consistency. All cost data given in this section has been converted into USD 2018 (inflation-adjusted).

2.1 High-temperature heat exchanger costs for powering the topping Brayton cycle externally

The most critical component of the proposed power plant concept (**Figure 1**) is the needed high-temperature gas–gas heat exchanger in order to power the topping Brayton cycle externally. As the coefficient of heat transfer on the atmospheric air side is very limited, the design is expected to be very bulky, since a large area of heat transfer is needed. In principal, a shell-and-tube heat exchanger design [11] is expected, having the pressurized air stream coming from the Brayton cycle's compressor on the tube side and the heating air stream at ambient pressure (coming from the TES) on the shell side. This type of heat exchanger will be similar to a heat recovery steam generator. Alternatively, and subject of this work, a regenerative heat exchange system working under atmospheric charging and pressurized discharging conditions can be applied [38] (see **Figure 2**), providing better heat exchange effectiveness. Clearly, the vessel size of this regenerative heat exchange system is limited due to the pressurization process, which requires several two-vessel subunits (such as shown in **Figure 2**) in parallel depending on the power rating. The second reason for several two-vessel subunits in parallel is the requirement for continuous thermal power transfer (while one system is pressurized/depressurized, the parallel systems need to take over). Thus, one disadvantage with respect to conventional heat exchangers is the higher complexity, as besides several parallel systems, high-temperature valves and piping are required for managing the pressurization/depressurization process. Furthermore, the pressurization process requires a certain amount of work, i.e., represents an additional parasitic consumption. This disadvantage needs to be offset by higher heat exchange effectiveness and reduced heat exchanger size (with respect to the conventional shell-and-tube layout). It is clear that this innovative regenerative system must have costs that are in the same order of magnitude as conventional heat exchanger designs in order to remain cost competitive. Here, cost figures published by Ilett and Lawn [39] are used. Calculating the cost difference between conventional combined cycle plants and externally fired ones results in a specific heat exchanger cost target of 64 kUSD per kg/s of air flow (topping Brayton cycle compressor air flow).

2.2 Gas turbine costs

The costs of the turbo machinery (compressor, turbine and turbo generator) are estimated as function of electric power output based on cost figures available in the open literature. Here it is important to capture the cost dependency on turbomachinery size, as smaller engines have higher specific costs (USD/kWe) than larger ones. Gas turbine costs are issued on a yearly basis by Gas Turbine World [40]. They propose a best fit curve, mentioning a $\pm$ 10% accuracy for gas turbine ratings ranging from 1 to 500 MWe. The investment cost of the turbo machinery IC_{GT} in USD (2018) is given as function of electric output power $P_{e\,GT}$ in kW in Eq. (2):

$$IC_{GT} = \left(9650 \cdot P_{eGT}^{-0.3}\right) \cdot P_{eGT} = 9650 \cdot P_{eGT}^{0.7} \tag{2}$$

2.3 Heat recovery steam generator (HRSG) costs

In order to obtain a reliable cost relationship for the heat recovery steam generator, the cost correlations published by Roosen et al. [41] and Silveira and Tuna [42] have been compared and agree very well once inflation-adjusted. Based on the correlation presented in Ref. [42], the following cost equation has been developed (in USD 2018), which only depends on heat duty $\dot{Q}$ (kW), air inlet temperature $T_{air\ in}$, stack temperature $T_{air\ out}$, and air mass flow $\dot{m}_{air}$ (kg/s):

$$IC_{HRSG} = 1.37 \cdot \left[4745 \cdot \left(\frac{\dot{Q}}{ln\left(T_{air\ in} - T_{air\ out}\right)}\right)^{0.8} + 2195 \cdot \dot{m}_{air}\right] \tag{3}$$

2.4 Steam turbine and remaining Rankine cycle component costs

Also steam turbine cost relationships published by Roosen et al. [41] and Silveira and Tuna [42] agree very well and, once inflation-adjusted, are also consistent with recent quotes requested by the authors. The adapted correlation (from Ref. [42]) is as follows:

$$IC_{ST} = 8220 \cdot P_{eST}^{0.7} \tag{4}$$

According to recent quotes, the cost estimate given by Eq. (4) may also include remaining Rankine cycle components, such as dry air-cooled condenser, feed water pumps, and deaerator. Finally, the above presented cost relationships have additionally been checked for consistency against Refs. [39, 43].

2.5 Heliostat field, tower, solar receiver, TES, and piping costs

Heliostat field costs are taken from Pfahl et al. [44], assuming 75 USD/m^2. This cost figure seems to be a realistic engineering target for heliostat designs that are optimized for mass production. The total heliostat field investment cost is obtained by multiplying the specific cost by the total solar field reflective area.

For the optimization process of the proposed multi-tower plant concept, small- to medium-sized fields are interesting, having tower heights in the range between 50 and 150 m. These tower heights are typical for wind turbines, and cost estimates for wind turbine towers should be applicable for solar power towers too, however considering larger tower diameters depending on needed piping diameter and receiver aperture size. Possible construction types for solar thermal power towers are either concrete type or metallic lattice type.

The following tower cost correlation has been established based on data given in Ref. [45] as function of tower height h_{tower} (m). The result, IC_{tower}, is the complete investment cost of the tower construction plus foundations and transport in M USD (2018), taking into account larger diameter towers providing enough space for the needed hot air piping as well as the receiver. Note that the valid height range for Eq. (5) is from 50 to 200 m:

$$IC_{tower} = 1.50227 - 0.00879597 \cdot h_{tower} + 0.000189709 \cdot h_{tower}^2 \tag{5}$$

The cost estimate of the solar receiver has been based on the CAPTure receiver prototype (≈ 300 kWth) costs, taking into account possible cost reductions when manufacturing the receiver up-scaled and in higher numbers commercially. The costs have been calculated per square meter of aperture area and result in 30 kUSD/m^2 for receiver aperture areas below 130 m^2, 50 kUSD/m^2 for receiver aperture areas until 400 m^2, and 100 kUSD/m^2 for bigger receiver aperture areas. The total aperture size of the receiver is an important factor as the receiver base structure (metallic + insulation) that supports modular ceramic absorber structures (cups + foams) as well as the air duct system becomes more complex and expensive the bigger the receiver is.

The costs of air/rock or air/ceramic thermocline packed-bed thermal energy storage can be assumed to be 20 USD/kWh_{th} [25]. However, in the case of the specific plant arrangement shown in **Figure 1**, the TES costs are higher since two low-temperature TES units are required for each high-temperature TES unit (regenerative use of return air heat). This approach is assumed to double the specific cost per kWh_{th}, resulting in 40 USD/kWh_{th} for the combined cycle option, only. Note that in the case of the Rankine single-cycle (benchmarking) configuration, the low-temperature TES units are not needed and the lower cost assumption applies.

The cost of internally insulated high-temperature piping has been assumed to be 800 USD per meter piping and per square meter flow cross section. It must be noted that the air speed in the air piping system must be kept reasonably low (≈ 20 m/s) in order to achieve acceptable pressure drop and thus blower power consumption.

The cost of blowers for circulation of air in the atmospheric circuit (blowers operate at ambient temperature) is assumed to be 3 kUSD per air volume flow (m^3/s). This cost assumption is based on several quotes requested by the authors.

Last but not the least, the yearly operations and maintenance (O&M) costs are assumed to be 1.5% of the total plant investment cost [3].

2.6 Levelized cost of electricity (LCOE) calculation

In the literature, the LCOE is an established figure for evaluating purely the economic lifetime energy production and its related lifetime costs, without taking into account revenues [46]. As revenues, i.e., feed-in tariffs, depend strongly on the country, the LCEO is therefore a relatively market-neutral figure and also allows to compare alternative technologies with different scales of investment or operating time [47]. Nevertheless, the LCOE depends on country-dependent parameters such as available solar resource, capital cost, and O&M costs, which must be taken into account in a serious technology benchmarking process. The general understanding of the LCOE in the literature [46, 48] is the total lifetime cost of the plant (engineering + construction + operation + maintenance + capital costs) divided by the lifetime electricity generation (total electric energy produced). Its unit is therefore cost per energy, i.e., USD/kWh. A particular point in the definition of the LCOE is that all costs incurred during the project lifetime are discounted back to the base year, i.e., their net present value (NPV) is taken into account [47]. Thus, according to Ref. [47], the LCOE can be calculated as given by Eq. (6). Note that C_n is the incurred cost in period n (engineering, construction, operation, maintenance, cost of capital), Q_n is the energy output in year n, d is the discount rate, and N is the total analysis period in years (power plant lifetime):

$$LCOE = \frac{\sum_{n=0}^{N} \frac{C_n}{(1+d)^n}}{\sum_{n=1}^{N} \frac{Q_n}{(1+d)^n}} \tag{6}$$

Also note that the applied discount rate should be the "real" discount rate, taking into account the inflation rate. A real discount rate of 3% is used in this work. The cost of capital for financing a CSP project is assumed to be 5% p.a. Power plant operating time is assumed to be 30 years (SEGS plants in the USA are in operation since the 1980s).

3. Power plant performance modeling

The power plant performance modeling is done as outlined in Ref. [10]. In particular, the solar receiver performance is estimated according to Ref. [49], using the detailed 1-D model to establish a receiver performance table as function of receiver operating temperature and incident solar flux. The topping Brayton cycle is modeled applying the isentropic relationships for air as ideal gas and choosing power class-dependent isentropic efficiencies. The bottoming Rankine cycle performance has been estimated applying state-of-the-art power cycle simulation software [43] and generating performance tables as function of HRSG inlet temperature and ambient temperature [10], suitable for annual yield simulations. The annual plant performance parameters (i.e., electricity yield, annual solar-to-electric efficiency) have been obtained running annual energy yield simulations using a typical meteorological year for Seville, Spain. The operating strategy is chosen such that the power block always operates under rated conditions (corresponding TES system charging/discharging) apart from start-up and shutdown periods.

3.1 Turbo machinery isentropic efficiencies as function of electric power output

As commonly known, the efficiency of turbomachinery is a function of power rating, i.e., the higher the output power, the higher is also the efficiency. Conversely, smaller engines have lower efficiencies. This is principally due to size-specific impacts of aerodynamic losses. For example, the turbine blade tip clearance (i.e., the radial distance between the blade tip of an axial compressor or turbine and the containment structure) is a major contributing factor to gas path sealing and can significantly affect engine efficiency [50]. The tip-leakage flow contributes negatively to the turbine performance and accounts for approximately one third of the total aerodynamic loss [51]. The bigger the engine, the smaller is the tip clearance with respect to the overall blade length and thus the higher is the efficiency.

It is clear that the size-dependent relationship of the turbomachinery's efficiency needs to be taken into account in the techno-economic optimization. In order to do so, relationships and performance tables have been established that consider efficiency as a function of output power, for both the topping Brayton cycle and for the bottoming Rankine cycle. For detailed information, the interested reader is referred to the corresponding public CAPTure project [38] deliverable D1.4 "CAPTure concept specifications and optimization."

4. Parametric benchmarking of power plant configurations: Combined cycle (CC) vs. Rankine single cycle (RC)

The principal objective of this section is to benchmark the techno-economic optimum of the CC plant against that of a conventional single-cycle Rankine steam plant with the same receiver and TES technology (see **Figure 3**). This will allow a fair assessment of the solar-powered combined cycle performance.

In order to analyze the impact of different solar field sizes and number of tower-solar-field modules, five solar field base modules (A, B, C, D, and E) have been selected (see **Table 1, Figures 4** and **5**). The applied solar field layout pattern is DELSOL [52], and solar field efficiency matrices can be obtained from CAPTure project deliverable D1.4. The base modules have been chosen such that different multiples achieve the same solar power class. For example, 9 B modules have the same nominal solar power as 3 C modules or 1 D module, i.e., 153 MW solar at the receiver(s). In this way, a direct comparison of conventional single-tower and multi-tower configurations can be achieved, giving also emphasis on the impact of total electric power of the plant. The general expected trends are that:

i. Smaller solar fields have higher optical efficiency.

ii. By arranging multiple solar field units as array, the optical efficiency for a given total solar power is improved; however, there is a point where HTF transport and additional tower and piping investment become too detrimental and the global performance is not better than that of a single-tower arrangement.

iii. Despite of much better optical efficiency of compact multi-tower arrangements, the considerable decrease in conversion efficiency of small power cycles, as well as elevated specific costs, generally makes small CSP plants economically unfeasible.

For each of the 19 configurations as indicated in **Table 1** (3 A to 9 A, 1 B to 9 B, 1 C to 9 C, 1 D to 4 D, and 1 E), the power plant performance models (combined cycle and single-cycle Rankine) have been run estimating the yearly energy yield and in consequence the resulting LCOE. The results are indicated in **Table 1** as well as in **Figure 6**. Note that the solar multiple (SM) and the TES capacity (hours of storage) have been optimized, i.e., obtaining the minimum LCOE at a solar multiple of about 2.3 and 10 full load hours of TES. The optimum solar multiple and TES capacity are typically only functions of geographic location and solar resource. When looking at **Table 1**, the first important trend that can be observed across all columns (all solar

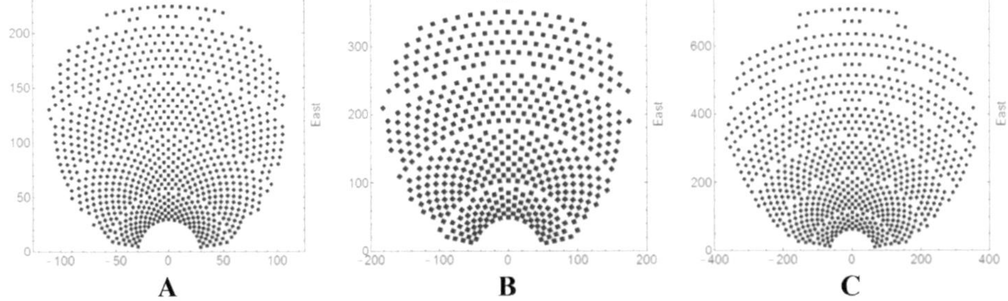

Figure 4.
Solar field base module types A, B, and C (solar field dimensions given in meters).

field base modules) is that when moving to a higher solar power class, the LCOE decreases. This is principally due to the fact that when moving to higher nominal power of the power block, the turbomachinery becomes more efficient and also the specific power block costs ($/kWe) reduce. This is the reason why commercial CSP projects have increased in size recently. However, when increasing the nominal solar power of a multi-tower arrangement (i.e., increasing the number of towers) above a certain threshold, the needed piping for HTF transport becomes an issue (investment, thermal losses, and pumping power); hence LCOE increases again (see configurations 9 C, 3 D, and 4 D). The second important trend that can be observed is that the single-tower configuration is only more competitive than a multi-tower configuration (of the same nominal solar power) below about 153 MW total nominal solar power. Above this threshold, the increase in investment (more towers, longer piping) and additional HTF transport power consumption are offset by the positive effect of higher optical efficiency of the more compact solar field base modules and smaller receiver aperture areas (cost advantage).

The most competitive (lowest LCOE) combined cycle power plant configuration is 6 C with a LCOE of 12.6 c$/kWh. However, the most competitive Rankine single-cycle plant configuration (also of type 6 C) achieves an LCOE of 11.5 c$/kWh. Thus, it is concluded that the combined cycle plant is despite its higher solar-to-electric conversion efficiency more expensive than the much simpler but less efficient single-cycle Rankine option. However, when observing **Table 1**, the difference in LCOE becomes smaller for smaller power classes, and the CC plant achieves better performance at 34 MW (and lower) total solar power. This is an effect of different power cycle efficiency decrease at small power classes, i.e., the combined cycle stays relatively more efficient than the Rankine single cycle configuration, which pays off for very small plant configurations. For this reason, the combined cycle seems to be

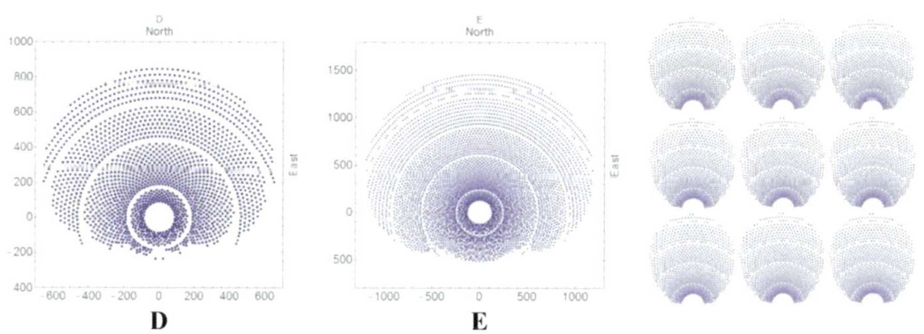

Figure 5.
Solar field base module types D and E (solar field dimensions given in meters). Multi-tower configuration 9 A (right-hand side).

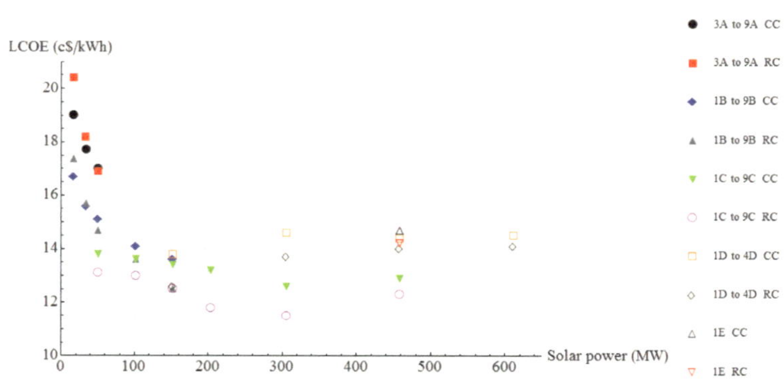

Figure 6.
LCOE as function of nominal solar power incident at receiver(s) and plant configuration (see also **Table 1***).*

Parameter (unit)	6C CC	6C RC	2B CC	2B RC
Number of towers (−)	6	6	2	2
Nominal solar power per tower (MW)	51	51	17	17
Total nominal solar power (MW)	306	306	34	34
Receiver thermal efficiency (−)/operating temperature (°C)	0.75/1050	0.81/ 800	0.75/1050	0.81/ 800
Solar-to-electric peak efficiency (−)	0.296	0.25	0.27	0.209
Solar-to-electric annual mean efficiency (−)	0.202	0.195	0.182	0.152
Solar multiple (−)	2.3	2.3	2.3	2.3
Power cycle nominal power (MWe)	50	45	4.9	3.7
Reheated GT nominal power (MWe)	28	—	3	—
Rankine cycle nominal power (MWe)	22	45	1.9	3.7
Power cycle annual mean conversion efficiency: CC/GT/RC (−)	0.496/0.288 / 0.355	—/—/ 0.388	0.434/0.268/ 0.277	—/—/ 0.295
TES thermal capacity (MWh)	981	1194	109	131.3
Yearly electricity yield (GWh)	161.6	156.8	15.6	13
Total plant cost (M USD)	175.35	154.55	20.93	17.51
Specific plant costs (USD/kWe)	3507	3434	4271	4732
Specific power cycle costs (USD/kWe)	849	678	1473	1423
LCOE (c$/kWh)	12.6	11.5	15.6	15.7

Table 4.
Power plant specifications of plant type 6 C and 2 B.

only attractive for very small power tower plants (below 5 MWe). Although gas turbines can be scaled down quite well having reasonable performance at small power classes, this is not the case for Rankine steam cycles. Hence, when thinking of very small (i.e., “micro”) combined cycles, the application of the organic Rankine cycle (ORC) as bottoming power cycle should be considered. This concept could be attractive for small and modular CSP central receiver plants for “electricity islands,” i.e., small remote grids, where electricity price is very high.

Table 4 shows the most important parameters of power plant configurations 6 C and 2 B. For plant configuration 6 C, it can be seen that although the combined cycle option achieves a higher solar-to-electric conversion efficiency, the increased plant complexity and thus its higher investment are not compensated by the increase in electricity yield. The combined cycle becomes cost competitive only at smaller power classes (see results for plant configuration 2 B in **Table 4**).

5. Conclusions

The parametric study shows that the multi-tower configuration has a techno-economic advantage with respect to the conventional single-tower arrangement above a total nominal solar power level of about 150 MW. The most competitive power plant configuration is of type 6 C. The combined cycle plant configuration reaches an LCOE of 12.6 c$/kWh, whereas the Rankine single-cycle power plant layout achieves 11.5 c$/kWh. Hence, the CC configuration has despite its higher solar-to-electric conversion efficiency a higher LCOE. The gain in electricity yield is not enough to outweigh the higher investment costs of the more complex CC plant layout. The CC configuration seems to be competitive only at smaller power classes. It must be said that all cost assumptions have inherent uncertainty, which makes a final conclusion regarding the best power plant layout very difficult. It is however clear that compact power plant arrangements (A, B, C options) are the preferred choice for the CAPTure power plant concept that applies atmospheric air as HTF, as large diameter piping (low air speeds are mandatory) becomes an issue at higher power classes, not only in terms of investment but also in terms of thermal inertia and losses. Therefore, it is very likely that in practical terms, a single-tower plant configuration will be the best choice when applying atmospheric air as HTF, as differences in LCOE are small. Furthermore, compact power tower plants have clear advantages regarding solar flux control, and also concerning total investment as financing is usually easier to obtain for smaller projects.

Finally, in order to make the CC attractive for CSP plants, the following challenges remain: (i) the efficiency of the solar receiver at relevant operating temperatures ($\approx 1000°C$) must be increased, and in particular innovative and economically competitive solar receiver designs are sought that allow long-term operation ($\approx$ 25–30 years) at very high solar flux densities, i.e. high concentration ratios; (ii) with regard to the investigated power plant layout, i.e. when using an open volumetric air receiver and atmospheric air as HTF, it is crucial to design a very economical high-temperature air–air heat exchanger train for powering the topping gas turbine externally.

Acknowledgements

This work has received funding from the European Union's Horizon 2020 research and innovation program under the grant agreement No 640905.

Author details

Fritz Zaversky[1*], Iñigo Les[1], Marcelino Sánchez[1], Benoît Valentin[2], Jean-Florian Brau[2], Frédéric Siros[2], Jonathon McGuire[3] and Flavien Berard[3]

1 National Renewable Energy Center (CENER), Solar Thermal Energy Department, Navarra, Spain

2 EDF – R&D, Chatou, France

3 Bluebox Energy Ltd., Hants, United Kingdom

*Address all correspondence to: fzaversky@cener.com

References

[1] Abbott D. Keeping the energy debate clean: How do we supply the world's energy needs? Proceedings of the IEEE. 2010;**98**:42-66

[2] ESTELA. The Value of Solar Thermal Electricity - Cost vs. Value Approach, ESTELA - European Solar Thermal Electricity Association, Brussels, Belgium; 2016

[3] Lilliestam J, Pitz-Paal R. Concentrating solar power for less than USD 0.07 per kWh: Finally the breakthrough? Renewable Energy Focus. 2018;**26**:17-21

[4] Ho CK. Advances in central receivers for concentrating solar applications. Solar Energy. 2017;**152**:38-56

[5] Ganapathy V. Industrial Boilers and Heat Recovery Steam Generators - Design, Applications, and Calculations. New York, USA: Marcel Dekker, Inc.; 2003

[6] Dunham MT, Iverson BD. High-efficiency thermodynamic power cycles for concentrated solar power systems. Renewable and Sustainable Energy Reviews. 2014;**30**:758-770

[7] Stein WH, Buck R. Advanced power cycles for concentrated solar power. Solar Energy. 2017;**152**:91-105

[8] Miller J. The combined cycle and variations that use HRSGs. In: Eriksen VL, editor. Heat Recovery Steam Generator Technology. Duxford, United Kingdom: Woodhead Publishing; 2017. pp. 17-43

[9] Fraidenraich N, Gordon JM, Tiba C. Optimization of gas-turbine combined cycles for solar energy and alternative-fuel power generation. Solar Energy. 1992;**48**:301-307

[10] Zaversky F, Les I, Sorbet P, Sanchez M, Valentin B, Brau J-F, et al. The challenge of solar powered combined cycles - providing dispatchability and increasing efficiency by integrating the open volumetric air receiver technology. Energy; 2020

[11] Al-attab KA, Zainal ZA. Externally fired gas turbine technology: A review. Applied Energy. 2015;**138**:474-487

[12] Siros F, Fernández Campos G. Optimisation of a low-TIT combined cycle gas turbine with application to new generation solar thermal power plants. In: ASME 2017 Turbo Expo. Charlotte, NC, USA; ASME; 2017. pp. V003T006A038

[13] Polyzakis AL, Koroneos C, Xydis G. Optimum gas turbine cycle for combined cycle power plant. Energy Conversion and Management. 2008;**49**: 551-563

[14] Becker M. GAST - The Gas Cooled Solar Tower Technology Program. Berlin Heidelberg, Germany: Springer Verlag; 1989

[15] Ávila-Marín AL. Volumetric receivers in solar thermal power plants with central receiver system technology: A review. Solar Energy. 2011;**85**:891-910

[16] Grange B, Dalet C, Falcoz Q, Ferrière A, Flamant G. Impact of thermal energy storage integration on the performance of a hybrid solar gas-turbine power plant. Applied Thermal Engineering. 2016;**105**:266-275

[17] Korzynietz R, Brioso JA, del Río A, Quero M, Gallas M, Uhlig R, et al. Solugas – comprehensive analysis of the solar hybrid Brayton plant. Solar Energy. 2016;**135**:578-589

[18] Puppe M, Giuliano S, Krüger M, Lammel O, Buck R, Boje S, et al. Hybrid high solar share gas turbine systems

with innovative gas turbine cycles. Energy Procedia. 2015;**69**:1393-1403

[19] Hennecke K, Schwarzbözl P, Alexopoulos S, Göttsche J, Hoffschmidt B, Beuter M, et al. Solar Power Tower Jülich - the First Test and Demonstration Plant for Open Volumetric Receiver Technology in Germany. Las Vegas, USA: SolarPACES; 2008

[20] Kwakernaak H, Tijssen P, Strijbos RCW. Optimal operation of blast furnace stoves. Automatica. 1970; **6**:33-40

[21] Hausen H. Berechnung der Wärmeübertragung in Regeneratoren bei zeitlich veränderlichem Mengenstrom. International Journal of Heat and Mass Transfer. 1970;**13**: 1753-1766

[22] Hennecke K, Hoffschmidt B, Koll G, Schwarzbözl P, Göttsche J, Beuter M, et al. The Solar Power Tower Jülich - a Solar Thermal Power Plant for Test and Demonstration of Air Receiver Technology. Beijing, China: ISES World Congress; 2007

[23] Zanganeh G, Pedretti A, Zavattoni S, Barbato M, Steinfeld A. Packed-bed thermal storage for concentrated solar power – Pilot-scale demonstration and industrial-scale design. Solar Energy. 2012;**86**: 3084-3098

[24] Zanganeh G, Pedretti A, Haselbacher A, Steinfeld A. Design of packed bed thermal energy storage systems for high-temperature industrial process heat. Applied Energy. 2015;**137**: 812-822

[25] ALACAES-SA, ALACAES - Creating Sustainable Energy Solutions for a Brighter Future, Lugano, Switzerland; 2017. Available from: https://alacaes.com/

[26] Fricker HW. Regenerative thermal storage in atmospheric air system solar power plants. Energy. 2004;**29**:871-881

[27] Hänchen M, Brückner S, Steinfeld A. High-temperature thermal storage using a packed bed of rocks - heat transfer analysis and experimental validation. Applied Thermal Engineering. 2011;**31**:1798-1806

[28] Moody LF. Friction factors for pipe flow. Transactions of the ASME. 1944; **66**:671-684

[29] Hoffschmidt B, Téllez FM, Valverde A, Fernández J, Fernández V. Performance evaluation of the 200-kWth HiTRec-II open volumetric air receiver. Journal of Solar Energy Engineering. 2003;**125**:87-94

[30] Téllez F. Thermal Performance Evaluation of the 200kWth "SolAir" Volumetric Solar Receiver. Madrid, Spain: CIEMAT-PSA; 2003

[31] Doligalski A, Sanchez de Leon L, Zachos PK, Pachidis V. Assessing the Potential of Gas-Recuperation in Reheated Gas Turbines Within Combined Gas-Steam Power Plants; 2014, V03AT07A031

[32] Sheikhbeigi B, Ghofrani MB. Thermodynamic and environmental consideration of advanced gas turbine cycles with reheat and recuperator. International Journal of Environmental Science and Technology. 2007;**4**: 253-262

[33] Sánchez M, Romero M. Methodology for generation of heliostat field layout in central receiver systems based on yearly normalized energy surfaces. Solar Energy. 2006;**80**:861-874

[34] Sment J, Ho CK, Moya AC, Ghanbari CM. Long-distance flux mapping using low-cost collimated pyranometers. Solar Energy. 2014;**100**: 76-83

[35] Ballestrín J, Marzo A. Solar radiation attenuation in solar tower plants. Solar Energy. 2012;**86**:388-392

[36] Sengupta M, Wagner M. Atmospheric Attenuation in Central Receiver Systems from DNI Measurements. Denver, Colorado, USA: ASES National Solar Conference; 2012

[37] Romero M, Marcos MAJ, Téllez FM, Blanco M, Fernández V, Baonza F, et al. Distributed power from solar tower systems: A MIUS approach. Solar Energy. 1999;**67**:249-264

[38] CENER, Horizon 2020 Research Project "CAPTure - Competitive SolAr Power Towers" - Grant Agreement Number 640905. 2015. Available from: www.capture-solar-energy.eu/

[39] Ilett T, Lawn CJ. Thermodynamic and economic analysis of advanced and externally fired gas turbine cycles. Proceedings of the Institution of Mechanical Engineers, Part A: Journal of Power and Energy. 2010;**224**:901-915

[40] Isles J. Gas Turbine World - 2018 GTW Handbook. Southport, CT, USA: Pequot Publishing Inc.; 2018

[41] Roosen P, Uhlenbruck S, Lucas K. Pareto optimization of a combined cycle power system as a decision support tool for trading off investment vs. operating costs. International Journal of Thermal Sciences. 2003;**42**:553-560

[42] Silveira JL, Tuna CE. Thermoeconomic analysis method for optimization of combined heat and power systems. Part I. Progress in Energy and Combustion Science. 2003; **29**:479-485

[43] Thermoflow-Inc., GT PRO - Gas turbine combined cycle design program to create cycle heat balance and physical equipment needed to realize it. 2018. Available from: http://www.thermof low.com/combinedcycle_GTP.html [Accessed: 27 July 2018]

[44] Pfahl A, Coventry J, Röger M, Wolfertstetter F, Vásquez-Arango JF, Gross F, et al. Progress in heliostat development. Solar Energy. 2017;**152**: 3-37

[45] Engström S, Lyrner T, Hassanzadeh M, Stalin T, Johansson J. Tall Towers for Large Wind Turbines, Elforsk - Electricity and Power Production. Stockholm, Sweden: Elforsk, Electricity and Power Production; 2010

[46] Cartelle Barros JJ, Lara Coira M, de la Cruz López MP, del Caño Gochi A. Probabilistic life-cycle cost analysis for renewable and non-renewable power plants. Energy. 2016;**112**:774-787

[47] Short W, Packey DJ, Holt T. A Manual for the Economic Evaluation of Energy Efficiency and Renewable Energy Technologies NREL/TP-462-5173. National Renewable Energy Laboratory; 1995. Available from: h ttps://www.nrel.gov/docs/legosti/old/ 5173.pdf

[48] IRENA, Renewable Energy technologies: Cost Analysis Series - Concentrating Solar Power, IRENA - International Renewable Energy Agency. 2012. Available from: www.ire na.org/Publications

[49] Zaversky F, Aldaz L, Sánchez M, Ávila-Marín AL, Roldán MI, Fernández-Reche J, et al. Numerical and experimental evaluation and optimization of ceramic foam as solar absorber – Single-layer vs multi-layer configurations. Applied Energy. 2018; **210**:351-375

[50] Chapman JW, Kratz J, Guo TH, Litt J. Integrated Turbine Tip Clearance and Gas Turbine Engine Simulation, NASA/TM—2016-219146. Cleveland, Ohio, USA: National Aeronautics and Space Administration - NASA; 2016

[51] Denton JD. The 1993 IGTI scholar lecture: Loss mechanisms in

Turbomachines. Journal of Turbomachinery. 1993;**115**:621-656

[52] Kistler BL. A User's Manual for DELSOL3: A Computer Code for Calculating the Optical Performance and Optimal System Design for Solar thermal Central Receiver Plants, Sandia National Laboratories, Albuquerque, New Mexico and Livermore, California; 1986

Chapter 3

The Emerging of Hydrovoltaic Materials as a Future Technology: A Case Study for China

Jiale Xie, Liuliu Wang, Xiaoying Chen, Pingping Yang, Fengkai Wu and Yuelong Huang

Abstract

Water contains tremendous energy in various forms, but very little of this energy has yet been harvested. Nanostructured materials can generate electricity by water-nanomaterial interaction, a phenomenon referred to as hydrovoltaic effect, which potentially extends the technical capability of water energy harvesting. In this chapter, starting by describing the fundamental principle of hydrovoltaic effect, including water-carbon interactions and fundamental mechanisms of harvesting water energy with nanostructured materials, experimental advances in generating electricity from water flows, waves, natural evaporation, and moisture are then reviewed. We further discuss potential applications of hydrovoltaic technologies, analyze main challenges in improving the energy conversion efficiency and scaling up the output power, and suggest prospects for developments of the emerging technology, especially in China.

Keywords: hydrovoltaic effect, carbon nanomaterial, electrokinetic effect, hydrovoltaic device, potential applications

1. Introduction

Water covers over 70% of the Earth's surface, which means it is abundant and widely available. Water contains enormous energy (35% of the solar energy received by the Earth, 10^{15} W) in a variety of forms, such as chemical, thermal, and kinetic energy [1]. The chemical energy is harnessed through water splitting under the assistance of electricity or photocatalysts [2]. The thermal energy is exploited for salinity power generation [3]. Kinetic energy is widely utilized by hydroelectric station, which is a main form of electricity.

As the progress of nanomaterials and nanotechnology, a new strategy based on hydrovoltaic effect (HV) has been developed in recent years [1]. In solar cells, the electron-hole pairs are generated by the absorption of photons with higher energy than the bandgap of semiconductor [4]. With the help of the built-in field at the interface of p-n junction, the electron-hole pairs are separated and then accumulated at the terminal of solar cells, generating photovoltaic voltage (**Figure 1a**). HV effect is analogous to the photovoltaic effect described as above. For HV effect, the potential is generated through the interaction between nanostructured materials and water molecules [1]. The form of water can be liquid, droplet, moisture, and

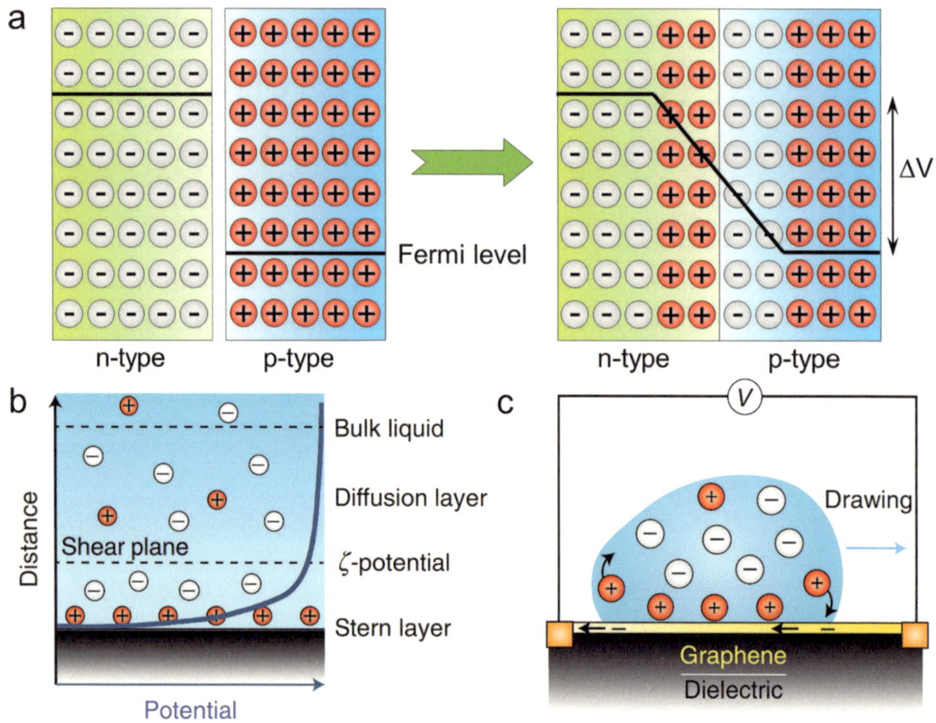

Figure 1.
(a) Photovoltaic effect with p-n junction. The basis of photovoltaic effect is the asymmetry of structural electronics. (b) Schematic of EDL forms at the solid surface with negative charges (not shown). There are two charge layers near the surface of solid, which are stern and diffusion layers. The stern layer is formed due to the chemical interaction between solid and absorbed ions. The diffusion layer electrically screen the stern layer through coulomb interaction. The blue line is the electrical potential curve around the solid surface [1].
(c) Illustration of induced potential by drawing a droplet on graphene. An electric current is formed in graphene by two moving boundaries of the EDL at the front and rear of the running droplet, respectively [1].

evaporation [5]. In brief, the basis of photovoltaic effect is the asymmetry of structural electronics (e.g., p-n junction). The nonuniformity of charge distribution at solid-liquid interface is the origin of HV effect. **Figure 1b** shows the electric double layer (EDL) at solid–liquid interface and the potential gradient as the distance increasing from solid surface into solution. **Figure 1c** illustrates the hydrovoltaic current is created in the graphene layer with the opposite orientation of the droplet flowing direction.

Since the HV phenomenon was discovered with carbon nanotubes (CNTs) in 2003, carbon nanomaterials were extensively investigated and considered as the most promising candidates for HV generators [6]. So far plenty of carbon nanomaterials exhibit HV effect with no need of a pressure gradient, including 0D graphene quantum dots (GQDs), 1D CNTs, 2D graphene or graphene oxide (GO), 3D graphene foam, and so on [7, 8]. Yet, unlike photovoltaic effect, research on the HV effect is in its infancy and calls for continued efforts to materialize its great potential. In this chapter, starting by describing fundamental principle of hydrovoltaic effect, including water-carbon interactions and basic mechanisms of harvesting water energy with nanostructured materials, experimental advances in generating electricity from water flows, waves, natural evaporation, and moisture are then reviewed. We further discuss potential device applications of hydrovoltaic technologies, analyze main challenges in improving the energy conversion efficiency and scaling up the output power, and suggest prospects for developments of the emerging technology.

2. Fundamentals of hydrovoltaic effect

2.1 Mechanisms of hydrovoltaic effect

2.1.1 Electric double layer and pseudocapacitance mechanism

As in **Figure 2a**, in a nanochannel, the EDL layers form on the interface of solid–liquid and overlap each other due to the small size of nanochannel. Under pressure gradient, a steady current will be generated along with the ion transport from high to low pressure side. The voltage will prohibit the transport of more ions. Therefore the steady current named the streaming current is generated [9]. There is positive correlation between the flow rate, pressure gradient, channel height, and the streaming current.

When a nanomaterial (e.g., graphene) is inserted into the liquid level, the EDL layer will be created and changed as the immersed area of nanomaterials changes. This means, if the immersed area increases, the EDL layer will be charged and a voltage will be generated in the nanomaterials (**Figure 2b**). On the contrary, an inverse voltage can be observed due to the discharging of EDL layer. This wave-induced voltage was called waving potential [10]. The voltage and current are proportional to the velocity of graphene and can be scaled up by series and parallel connections of multiple graphene devices.

When the water contained ions is a droplet on graphene, the EDL is emerged only in the region of droplet followed the EDL theory. When the droplet is moving under the external force such as gravity, the EDL region will move accordingly. During this moving process, there is a charging state at the front of the droplet, while a discharging process at the rear of the droplet (**Figure 1c**). Therefore an electrical voltage called as drawing potential can be generated [11]. The voltage and current will increase as the velocity and number of droplets increase. The drawing potential can be developed to harvest raindrop energy (**Figure 2c**).

2.1.2 Ion diffusion-induced mechanism

When moisture were adsorbed by nanomaterials with oxygen-containing functional groups, a gradient of H^+ can form because of the local salvation effects, which can lead to the breakage of $O^{\delta-}$-$H^{\delta+}$ bonds. Due to the H^+ concentration difference, the H^+ will migrate along the reverse gradient direction. Then a voltage would increase continuously until the gradient of H^+ vanishes. When the ingress of moisture stopped, the number of migratory ions decreased as free H^+ and oxygen-containing functional groups recombined, resulting in an reverse voltage [12].

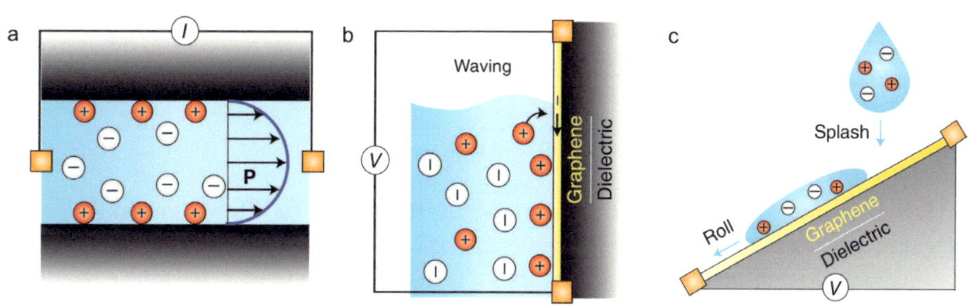

Figure 2.
(a) Schematic of electrokinetic effect in the nanochannel. Blue line illustrates the profile of flow velocity through the nanochannel [1]. (b) Illustration of waving potential induced in graphene by one moving boundary of EDL across a graphene sheet on a dielectric substrate [1]. (c) Schematic illustration of harvesting energy from raindrops [1].

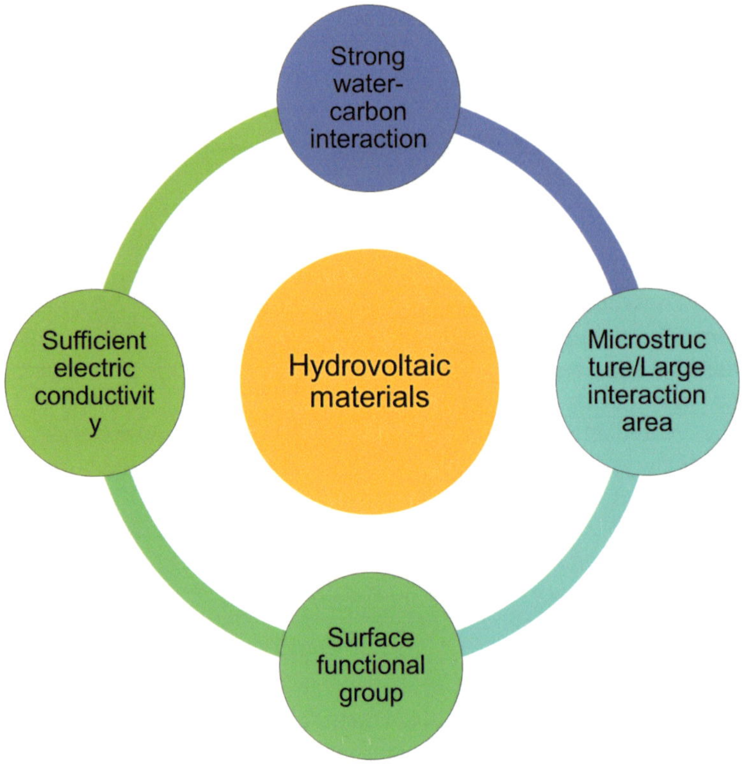

Figure 3.
Characteristics of a good hydrovoltaic material.

2.2 Rules for hydrovoltaic material

Based on the above discussion of mechanisms, an excellent hydrovoltaic material should possess the following characteristics: strong water-carbon interaction, rational pore and microstructure for the transport of water molecules, large interaction area for water adsorption and charge storage, sufficient electric conductivity for charge transportation, and abundant/gradient surface functional groups (oxygen functional groups in particular). Nowadays, the most reported hydrovoltaic materials are carbon nanomaterials [7, 8]. However, the hydrovoltaic effect is not limited to carbon nanomaterials but can be generic to other materials as long as they meet the above characteristics (**Figure 3**).

3. Hydrovoltaic devices and performance

3.1 Moisture-induced electricity generation

Moisture is one important form of water in the nature. GQDs have unique properties due to quantum confinements and edge effects [13]. GQDs as the chemically active material have fabricated a moisture-triggered generator [14]. The size of GQDs is 2–5 nm. The GQDs contain an amount of oxygen-containing functional groups. To create a gradient of functional groups, GQDs are treated via electrochemical polarization. The GQD generator achieves a high voltage of 0.27 V, when the variation of relative humidity (RH) is 70%. After the optimization of the

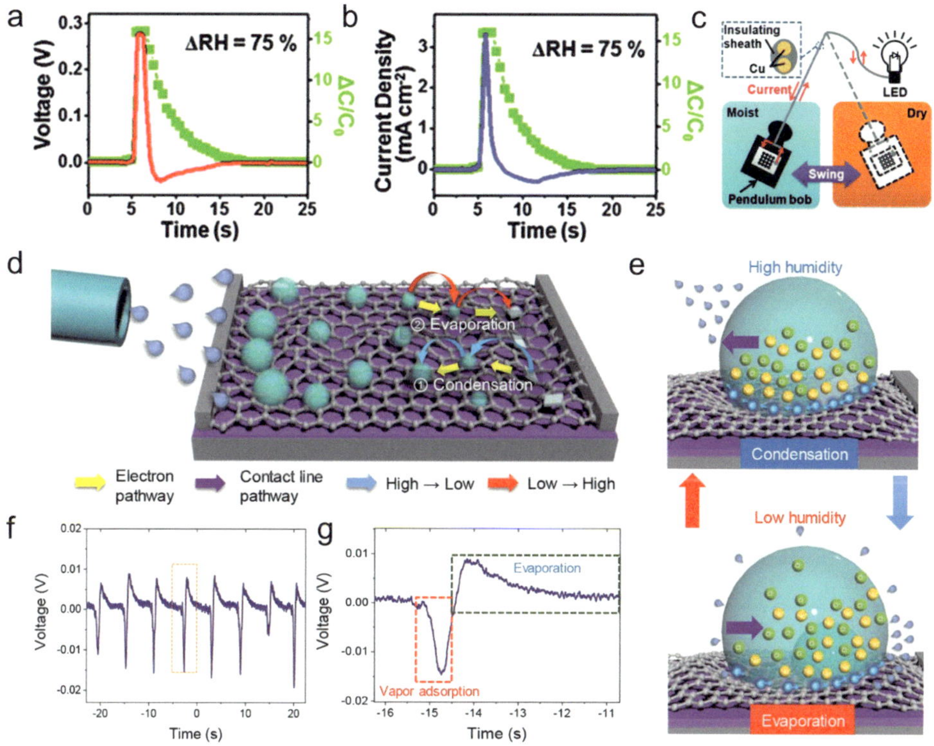

Figure 4.
(a) Voltage and (b) current output cycle of HV device based on g-3D-GO that is sandwiched by aluminum electrodes in response to the RH variation (ΔRH = 75%) [16]. (c) Schematic illustration of a HV device-based power source system [16]. (d) Schematic illustration of the mechanism of humidity-driven electricity generation [17]. (e) Condensation and evaporation of ionic liquids under different humidity [17]. (f and g) Voltage generated with wrinkle graphene/salt crystal nanogenerator under a sudden change in humidity [17].

load resistor, a power density obtained is 1.86 mW/cm^2. The gradient of oxygen-containing functional groups is the reason of electricity generation with moisture. Similarly, the porous carbon black, and GO framework with the functional group gradient, can also exhibit excellent HV performance under moisture [15, 16]. For example, a superhydrophilic 3D assembly of graphene oxide (g-3D-GO) with open framework exhibits a high power density of ca. 1 mW/cm^2 and an energy conversion efficiency of ca. 52% [16]. With an RH variation of 75%, the g-3D-GO-based HV device could provide a voltage and current output of ca. 0.26 V and ca. 3.2 mA/cm^2 within 2 s (**Figure 4a** and **b**). As in **Figure 4c**, a power source system consists of four HV cells in series which was fabricated to demonstrate the practical application. This system was attached onto the pendulum bob. The pendulum bob can swing between moist region (RH = 80%) and dry region (RH = 5%). When the system moves to the moist region, the moisture-induced positive voltage is applied on the light emitting diode (LED), and it lights up. Upon traveling to the dry region, the LED will switch off. As a consequence, this power source system could provide a steady power output.

Recently, Zhen et al. prepared a nanogenerator using the wrinkled graphene, which followed an unusual mechanism of HV effect [17]. In this work, a new cation-π interaction utilization strategy was developed. In other words, electricity is generated through water adsorption and desorption of salt crystals along with the humidity variation. The key of this nanogenerator is to deposit salt crystals onto the wrinkled graphene by manipulating the formation of ionic liquid microdroplets.

The wrinkled graphene has many defects and uniform wrinkles, facilitating the ultrafast water evaporation, preventing excessive water accumulation and deposition of well-distributed salt crystals. **Figure 4d** and **e** schematically illustrates the mechanism of electricity generation. As the sudden change of humidity (25–75–25%), two inversed voltage peaks were observed sequentially (**Figure 4f**). This is attributed to the water vapor adsorption and desorption on the salt crystals. The sharper negative peak is due to the strong water adsorption ability of the salt crystals, while the broad positive peak is from the slow desorption process (**Figure 4g**). The voltage of 18 mV with the current of 37 nA was achieved with a $1 \times 6\ cm^2$ generator. Among various salts, NaCl exhibits the best performance due to its complete crystallization after each cycle.

As far as we know, the nanomaterials for moisture-induced electricity generation include carbon/graphene quantum dots, carbon black, GO film, and 3D GO frameworks. The main origin of electricity generation is similar to each other. The potential generation is dependent on the water adsorption difference due to the gradient of oxygen-containing functional groups and induced the concentration difference of charge carriers. Interestingly, the porous carbon black film treated partially by plasma could generate continuous electricity, which is totally different from other carbon nanomaterials. This discrepancy may be from the difference of the structure and/or the introducing manner of functional groups.

3.2 Electricity generation induced by droplet movement

In 2014, Yin et al. firstly reported the electricity generation induced by droplet movement on monolayer graphene [11]. When a droplet of 0.6 M NaCl is drawn on graphene at a constant velocity of 2.25 cm/s, a voltage of 0.15 mV is generated. When the direction of droplet movement is opposite, the direction of potential is also reversed. When the movement is stopped, no potential is produced. Rain is one of important existent forms of water in nature. However, the energy in the rain is not yet utilized efficiently in the long term. There are amount of cations (such as Na^+, NH_4^+, Ca^{2+}, Mg^{2+}) and anions (such as Cl^-, NO_3^-, SO_4^{2-}). Therefore harvesting energy from rain using HV effects is a promising approach [18, 19].

As previously reported, the low generated voltage of around 0.15 mV, and the external pressure needed, would limit the application of HV effect. Recently, Li et al. reported the electricity generation from water droplets on porous carbon film through capillary infiltrating [20]. **Figure 5a** illustrates the structure of the porous carbon film (PCF) device. When a droplet of 1 μL was dropped onto PCF, a sustainable voltage of 0.3 V was generated (**Figure 5b**). However, electricity generation by droplet movement on graphene or aligned single-walled nanotubes is pulse-like. The retention of voltage depends on the volume of water droplets as shown in **Figure 5c**, but the generated voltage value is nearly identical. More interesting, the dropping position of water droplet would influence the induced voltage (**Figure 5d**). Experimental results reveal the following key characteristics: (i) the merely directional water infiltration can induce the voltage, (ii) no direct correlation between the HV voltage and the position of droplets, and (iii) the direction of water infiltration influences the voltage sign. At last, the authors demonstrated a scale-up application with three devices in series (**Figure 5e**). Twelve 5 μL water droplets can generate a voltage up to 5.2 V and illuminate a liquid crystal display. This work powerfully demonstrate that a hydrophilic porous carbon film with water droplets could realize the energy harvested from rain and a practical application. However, there is no experimental results in real raining environment. More information on the droplet-induced electricity generation can refer one recent review paper [21].

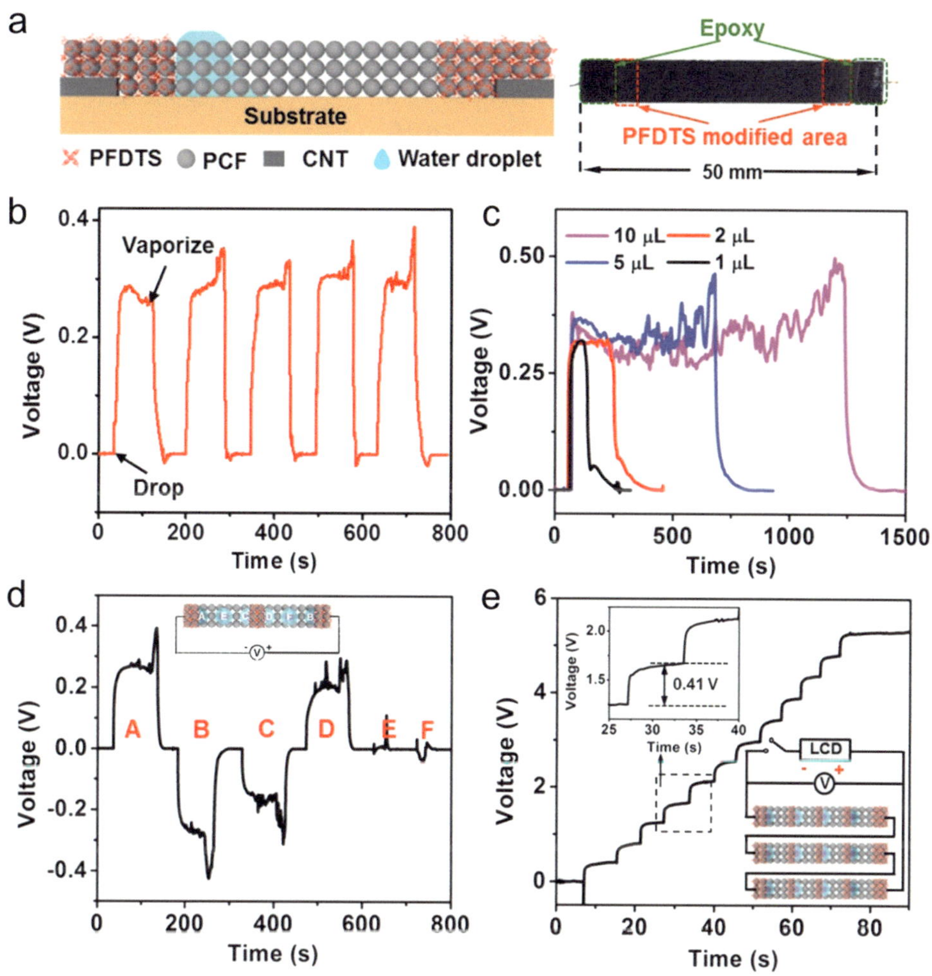

Figure 5.
(a) Schematic of the porous carbon film device with two ends modified with 1H,1H,2H,2H-perfluorodecyltriethoxysilane (PFDTS). The right inset shows a photograph of a typical device with dimensions of 50 × 7 mm² [20]. (b) Open-circuit voltage obtained by repeatedly dropping 1 μL water droplets at the PFDTS@PCF/PCF interface under ambient conditions (~ 23.5°C and RH ~ 71.7%) [20]. (c) Measured V_{oc} vs. time of the device when water droplets with various volumes were dropped onto the PFDTS@PCF/PCF interface [20]. (d) Wetting dependence of the induced voltage. Inset is schematic of the V_{oc} measurement and the water-droplet position [20]. (e) Application demonstration of the water-droplet-induced voltage [20].

3.3 Flow-induced electricity generation

Ocean wave energy is a main form of ocean energy, which is considered as inexhaustible energy. In 2007, Liu and Dai reported that the flow-induced voltage can be greatly improved by aligning the nanotubes along the flow direction [23]. In 2017, Xu et al. fabricated a fluidic nanogenerator fiber with the aligned multi-walled carbon nanotube sheet (inset of **Figure 6b**) [22]. The device shows a power conversion efficiency of 23.3% and an excellent stability over 1,000,000 cycles. The flow direction, the flow distance, the flow velocity, and the NaCl concentration are positive correlation with the induced voltage (**Figure 6a–6c**). The authors also discovered that the ordered mesoporous carbon (OMC) can significantly enhance the flow-induced voltage (**Figure 6d**). After OMC introduction, the sustained voltage for over 1 h can be achieved. The maximum voltage output can reach up to 341 mV when the content of OMC is 5.1 μg/cm. Impressively, the stable performance of this

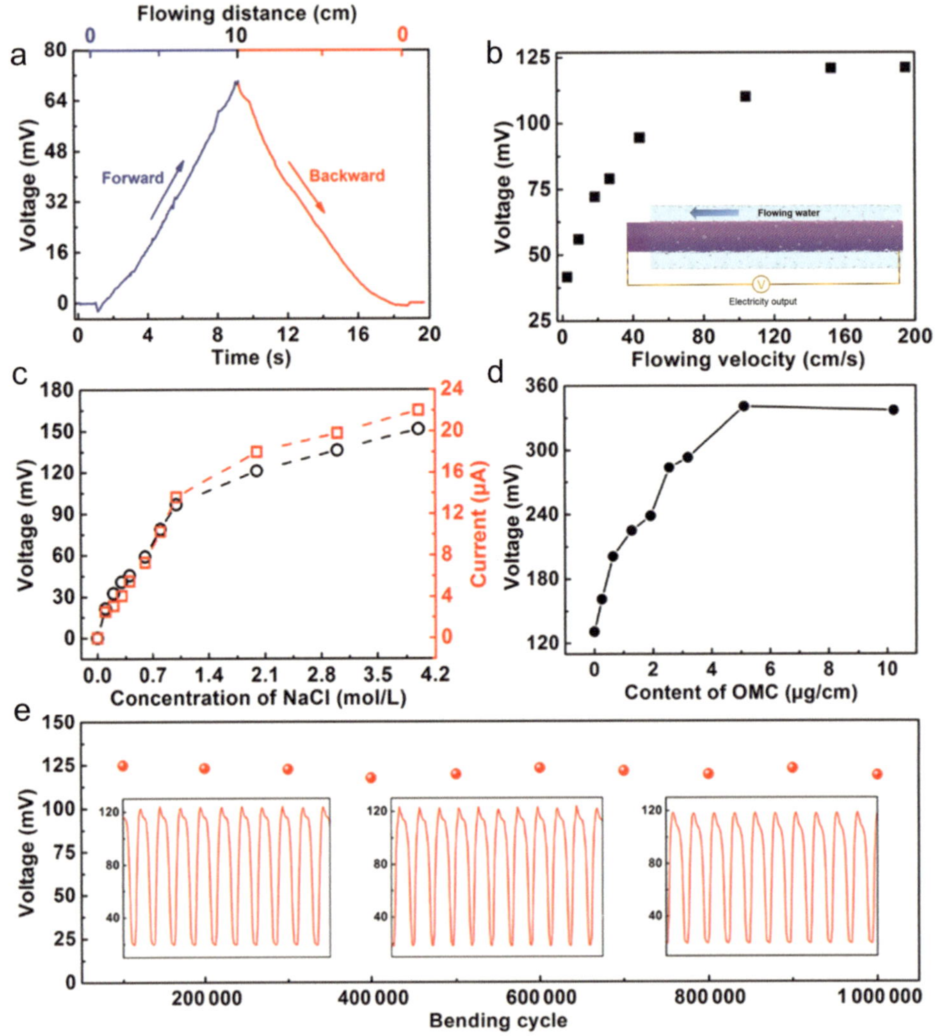

Figure 6.

(a) Voltage curve induced by a saturated NaCl flow at the velocity of 1.2 cm/s [22]. (b) Relationship between the voltage and the flow velocity. Solution is 0.6 M NaCl [22]. (c) The voltage vs. current relationship as the concentration variation of NaCl solution (flow velocity: 12.9 cm/s) [22]. (d) Dependence of the voltage on the OMC content in a saturated NaCl solution (flowing velocity: 20 cm/s) [22]. (e) Voltage generated by repeatedly dipping an OMC-incorporated device into a NaCl solution with an increasing number of bending cycles. The inserted graphs show the voltages after 200,000; 600,000; and 1,000,000 bending cycles in the NaCl solution [22].

device can be maintained even after over 1,000,000 bending cycles (**Figure 6e**). Moreover, the fiber nanogenerator is flexible and stretchable, indicating it can be woven into fabrics for large-scale applications.

Two-dimensional materials or devices have more advantages for ocean wave energy harvesting, which can well float on the surface of the ocean [1, 25]. Recently, Fei et al. achieved volt leveled waving potential using a pair of graphene sheets [24]. **Figure 7a** illustrates the device setup, where a pair of graphene-PET sheets with size of 2.5 × 1.5 cm^2 is immersed in NaCl solution. As one of the graphene sheets moves through the liquid surface, electricity is generated (**Figure 7b**). The peak voltage can be around 60–120 mV. However, no voltage is observed either moving graphene underneath or parallel to the liquid level, indicating the waving potential is due to the dynamic EDL boundary. In this setup, the moving graphene is served as driving force for ion movement, while another graphene is as a reference electrode.

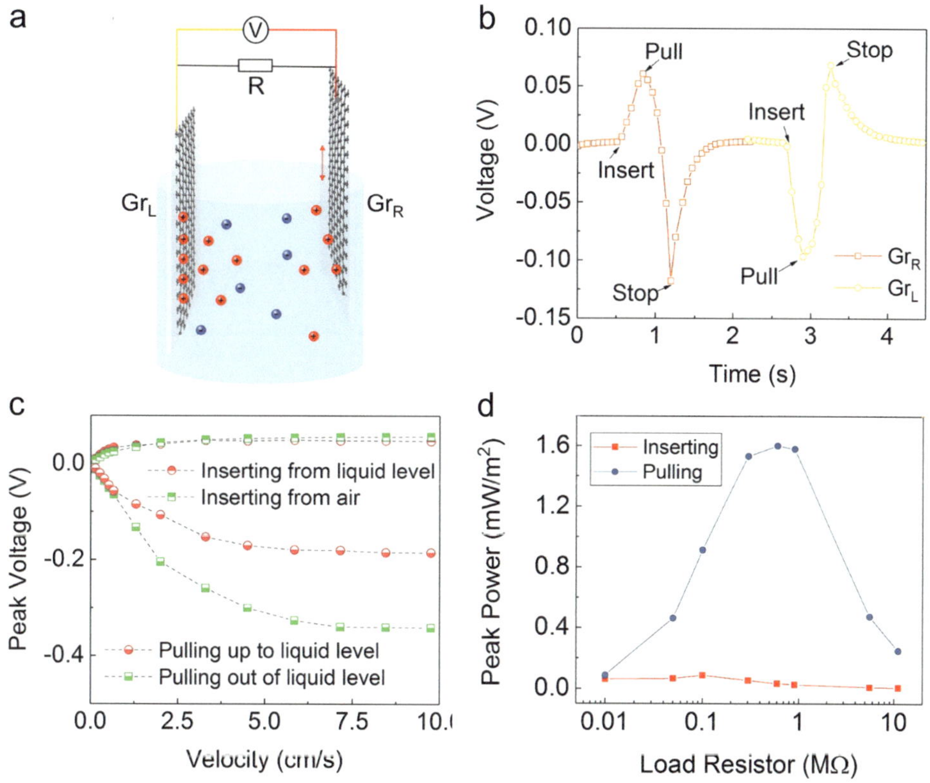

Figure 7.
(a) Schematic of experimental setup with two graphene-PET sheets immersed vertically in an electrolyte container. Gr_L *and* Gr_R *represent graphene samples on the left and right side, respectively [24]. (b) Generated voltage signals at resistor of 0.9 MΩ when separately moving* Gr_L *or* Gr_R *across NaCl solution [24]. (c) Peak voltage values collected as a function of velocity. The perpendicular moving distance of graphene is kept as 2 cm [24]. (d) Calculated output power per unit area of graphene [24].*

Figure 7c shows the relationship between the peak voltage and the moving velocity of graphene. The peak voltage exhibits linear relations with velocity at low moving speeds and saturates to certain values at high speeds. This saturation may be due to the limit from the speed of ion adsorption/desorption. During the pulling process, when the load resistance is 0.6 MΩ, the largest power density of 1.6 mW/m^2 can be obtained (**Figure 7d**). Moreover, the ion species can also influence the voltage value. The peak voltage values follow an order of LiCl > NaCl > KCl > $BaCl_2$, indicating the smaller ions are better for higher voltage generation.

Apart from the dynamic EDL boundary mechanism, the water-carbon interaction can also generate electricity [26]. In this case, the water does not need the cations and anions for the formation of EDL capacitance. More importantly, the electric signals are continuous, but the inherent mechanism is not very clear yet. More information on the liquid flow-induced electricity in carbon nanomaterials can refer the recent review papers [27, 28].

3.4 Evaporation-induced electricity generation

Water evaporation is a crucial step in the natural water circulation, releasing a huge amount of water energy. In 2017, Xue et al. reported that the water evaporation from centimeter-sized carbon black sheets can reliably generate sustained voltages of up to 1 V for 8 days under ambient conditions [29]. The annealing and plasma treatment introduced functional groups are essential for the electricity generation.

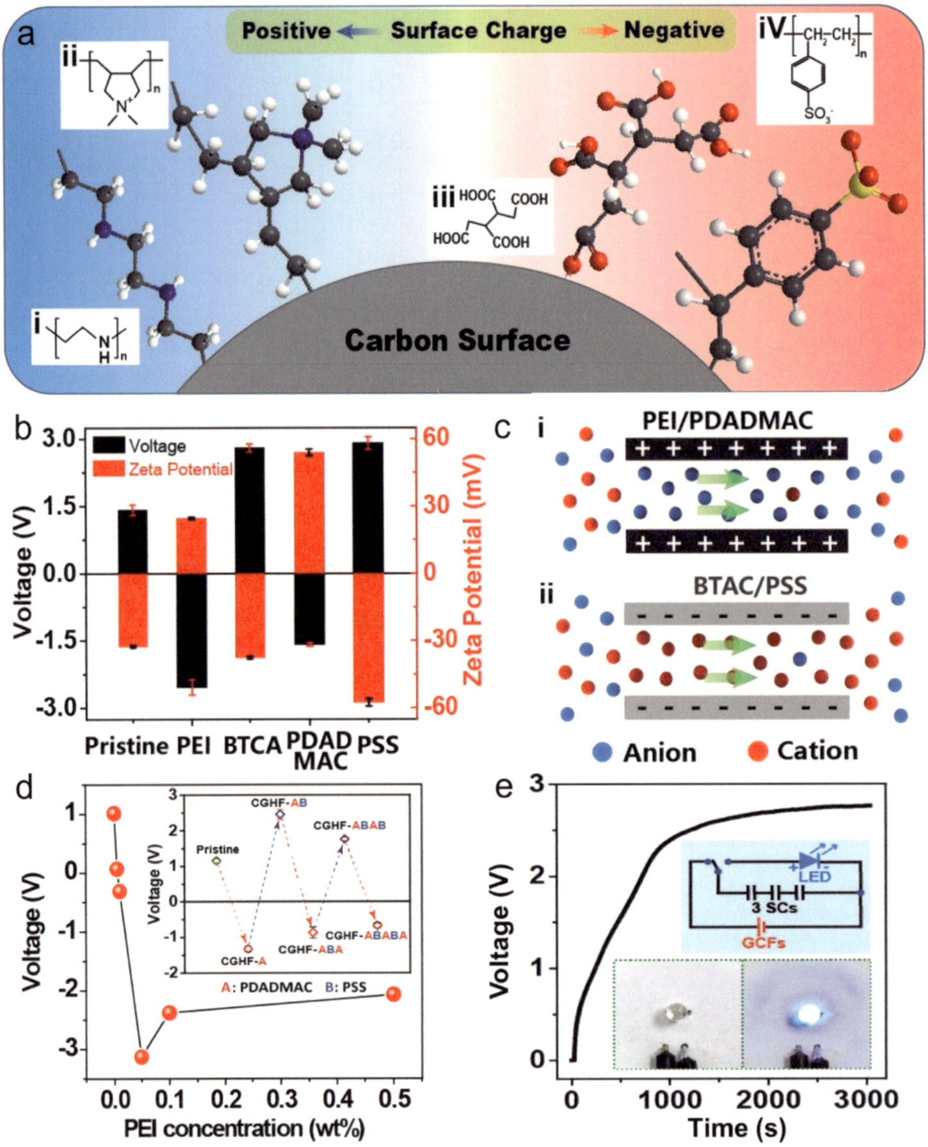

Figure 8.

(a) Schematic of the chemical modification with different molecules on carbon nanoparticles [30]. (b) V_{oc} and zeta potentials of the pristine and modified GCFs [30]. (c) Schematic of the ion-selective transport mechanism in the nanochannels with (i) positively and (ii) negatively charged surface [30]. (d) Dependence of V_{oc} of the device on the PEI concentration. The inset shows the evolution of V_{oc} when a GCF was repeatedly modified by DADMAC(A) and PSS(B) [30]. (e) Voltage-time curve of three supercapacitors (SCs) connected in series in charging by the GCFs at ambient condition. Insets show the circuit diagram (top) and photos of the blue LED [30].

Recently, Li et al. fabricated an evaporation-driven nanogenerator (1 cm × 5 cm) with a high open-circuit voltage of 3 V [30]. The film device is fabricated using carbon black and glass fiber. As in **Figure 8a**, the surface of the hybrid film was modified with several polymer molecules, such as polyethylene imine (PEI); 1,2,3,4-butane tetracarboxylic acid (BTCA); polydimethyl diallyl ammonium chloride (PDADMAC); and poly sodium-p-styrenesulfonate (PSS). The voltage of the glass-fiber-carbon-nanoparticle film (GCF) can vary from −3 V to 3 V, which rely on the difference of the surface functional groups (surface charges) as in **Figure 8b**. PEI- and PDADMAC-modified GCF have positive surface charges and thus positive zeta potentials. The ion-selective transport mechanism is schematically shown in **Figure 8c**. The concentration of polymer solution is one of critically experimental conditions, which can significantly affect the voltage from +1 V to −3.2 V by using 0–0.5 wt% PEI solution (**Figure 8d**). There is an optimal concentration of PEI solution (0.05 wt%), which should be attributed to the low conductivity of polymer and/or the partial channel blocking. A hybrid film device using two GCFs with opposite surface charges was also prepared. Therefore the generated voltage was enhanced to around 5 V (5 × 5 cm). They also connected the hybrid film device with supercapacitor. The supercapacitor can store the electric energy from GCFs and provide a high current output. The supercapacitor can be charged up to 2.8 V by the output voltage of GCFs, and then a blue LED can be lighted up for about 10 s without any auxiliary. This work shows the great potential of evaporation-induced electricity generation in the field of portable electronics.

4. Potential applications

4.1 Self-powered liquid sensors

Because of the low power density (10^{-3}–10 W/m^2) of HV device, [1] it is a good choice to develop HV effect-based self-powered devices, such as self-powered liquid sensors. The factors which can affect the HV signal could be used for sensing applications, including flow rate, fluid volume, solution component, fluid movement, and so on. Yin et al. developed a monolayer graphene-based HV device in 2013 and demonstrated the self-powered liquid sensor application [11]. The configuration of HV devices is shown in **Figure 9a**. A droplet of 0.6 M NaCl was moved with a SiO_2/Si wafer on the graphene film. During the moving of the droplet, the advancing and receding contact angles are ~91.9° and ~60.2°, respectively. As in **Figure 9b**, the

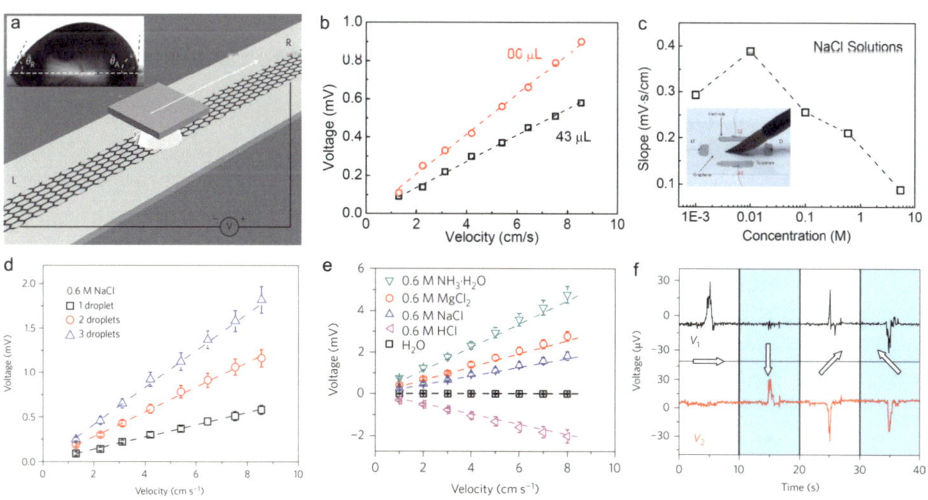

Figure 9.
(a) A liquid droplet is sandwiched between graphene and a SiO_2/Si wafer and drawn by the wafer at specific velocities. Inset: A droplet of 0.6 M NaCl on a graphene surface [11]. (b) Dependence of the output voltage on the volume of a droplet of 0.6 M NaCl [11]. (c) Dependence of the output voltage on the concentration of the solution (three droplets of NaCl solution). Inset: Photograph of handwriting with a Chinese brush on graphene [11]. (d) Voltage induced by moving one, two, and three droplets of 0.6 M NaCl. Dashed lines are curves linearly fitted to the measured data [11]. (e) Voltage for various ionic solutions (three droplets) on monolayer graphene [11]. (f) Sensing the stroke directions (arrows) by the drawing potentials between electrodes E_1^+–E_1^- and E_2^+–E_2^- as shown in the inset of c [11].

voltage of device is linearly proportional to the velocity of droplet. The larger the size of the droplet would induce the larger voltage. The concentration of NaCl solution is also a critical factor, but the trend is not monotonic. The best concentration of NaCl solution is around 0.01 M. The voltage output is less than 0.5 mV. However, in carbon black, the potential is maximum ~1.0 V when deionized (DI) water is used for evaporation HV generator [29]. It is interesting that the voltage can be multiplied by drawing multiple droplets simultaneously as shown in **Figure 9c**. Experimental results indicate the voltage is near zero when two droplets are moving in the opposite directions. The voltage of device is closely related to the ion species, such as $MgCl_2$, HCl, and $NH_3{\cdot}H_2O$. However, there is no response for DI water. This indicates the drawing potential comes from the ion-induced EDL capacitance. The voltage induced with HCl solution is negative. The authors think that there is a H_3O^+ layer on the surface of graphene. Therefore the positively charged graphene would attract the negative Cl^- anions, which dominate the electric double layer. At last the authors show a handwriting sensor with a Chinese brush and 0.01 M NaCl. Two pairs of electrodes, E_1^+–E_1^- (right–left) and E_2^+–E_2^- (bottom-top), were patterned perpendicularly on the four sides of the graphene to distinguish the handwriting direction (inset of **Figure 9c**). The drawing direction related to voltage signal can be well detected as in **Figure 9f**. Moreover, the force and speed of the handwriting can also be monitored.

Graphene oxide framework with lots of pores can facilitate the diffusion of water molecules. Along with the asymmetrical oxygen-containing groups, Zhao et al. observed that the transport of the ionic charge carriers is accelerated due to the ionic gradient [16]. When the RH variation is 75%, the potential can increase up to 260 mV in 2 s. The concentration gradient of Li ions in 3D PPy framework can also show good sensitivity for moisture [31]. The potential is 60 mV under the RH variation of 85%. More importantly, the stability of this material is stable during the several hundred cycles.

Besides the HV potential, other signals can also be utilized for sensing, such as the resistance of materials, the length of fibers, and the volume of materials [32–36]. Of course, the change of other physical fields, such as temperature and wind speed, can be detected by HV devices directly or indirectly [29, 37].

4.2 All-weather power generation

As we all know, the common solar cells only work on sunny days, but do not work in the night and on rainy days. Combining PV and HV effects, a hybrid cell for all-weather power generation was developed by integrating a solar cell with a HV device. Tang et al. fabricated a flexible solar cell made of a transparent graphene electrode and a dye-sensitized solar cell [38]. The hybrid cell can be excited by solar light on sunny days and raindrops on rainy days, yielding a solar-to-electric conversion efficiency of 6.53% under AM 1.5 irradiation and power in the range of 5.12–54.19 pW by simulated raindrops (**Figure 10b**). However, the output of this hybrid cell is still far lower than the actual requirement. Then Tang et al. changed the graphene to graphene/carbon black/polytetrafluorethylene (PTFE) for the hybrid cell fabrication [18]. But the voltage and current did not show significant improvement under simulated raindrops. Zhong et al. developed a 2D hybrid nanogenerator based on graphene and silicon for all-weather electricity generation [39]. In this hybrid cell, the graphene and silicon form the van der Waals Schottky diode (**Figure 10c**). This hybrid device delivers a maximum output power of 49.3 μW under light illumination. When the DI water flows on the graphene under light from Au electrode to Ag electrode, an additional potential of 2.54 mV can be generated (**Figure 10d**). However, there is no response in the dark, indicating no interaction

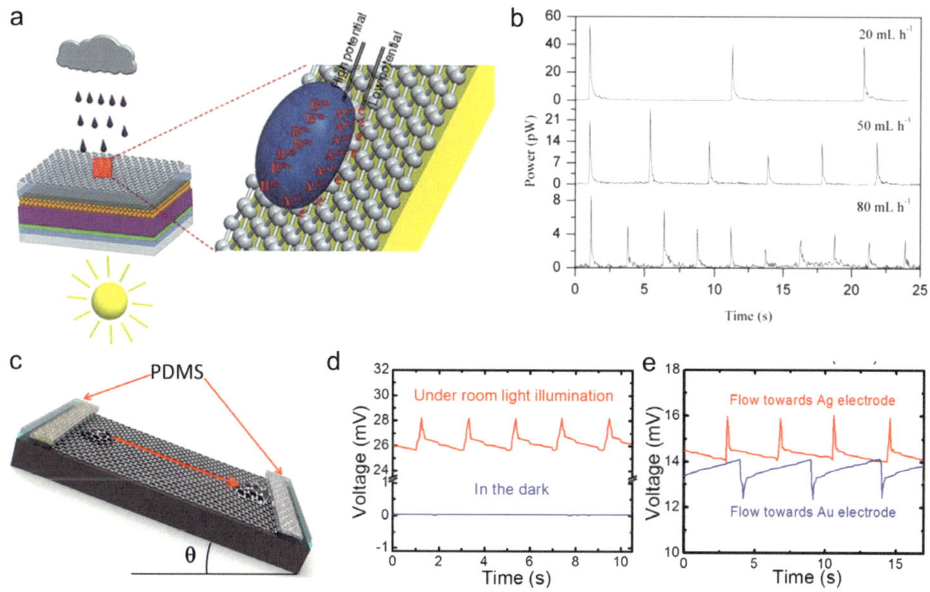

Figure 10.
(a) The quasi-all-weather solar cell that can produce electricity from rain and sun [38]. (b) Power signals produced by dropping 0.6 M NaCl droplets on rGO film [38]. (c) Schematic structure of the 2D hybrid nanogenerator of graphene/SiO_2/Si [θ = 30°, PDMS: Poly(dimethysiloxane)] [18]. (d) Voltage responses to the flow of DI water over graphene/Si Schottky diode in the dark and under room light illumination [18]. (e) Voltage responses to the flow of DI water toward different directions [18].

exists between water and hybrid cell during the water flowing process. The authors think that the potential response should arise from the interaction between water and graphene/Si Schottky diode (the doping and dedoping at the front and rear side of water droplet, respectively) instead of the water-graphene interaction or the water-electrode interaction. Moreover, the negatively additional potential can be observed when the water flows toward Au electrode. This phenomenon should be attributed to the asymmetric potential profile of the graphene channel.

4.3 Harvesting Ocean wave energy

Due to the intermittency and randomness of raining, it is wise to harvest energy from the ocean wave energy. The ocean wave energy is renewable and inexhaustible because 70% of the Earth's surface is covered by the ocean. The ocean wave energy is also called as blue energy [40]. Tan et al. fabricated a film type wave energy generator with graphene, carbon black, and polyurethane [25]. A voltage of > 20 mV and a current of > 10 μA are produced in a 15 cm^2-sized generator. Moreover, the devices are sustainably stable upon persistent attack by waving ocean. The floating devices on the sea can be packaged into the wave energy stations by series and parallel connections. Tan et al. further proposed a photo-induced charge boosting liquid-solid electrokinetic generator with a structure of polyurethane/graphene oxide-carbon black-multi-walled carbon nanotube/carbon quantum dots/copper (PU/GO-CB-MWCNT/CQDs/Cu) [41]. Under AM1.5 illumination, the voltage, current, and power density achieved by this device are 0.1 V, 0.39 mA, and 26.6 mW/m^2, respectively. The working mechanism is described in **Figure 11a-c**. Due to the difference of Na^+ and Cl^- ions on adsorption energy, the EDL can be formed. Same as the principle of waving potential as above, the potential signal would change along with the seawater moving and the capacity change. **Figure 11d** presents an enlarged one-cycle electrical signal generation. When the seawater moves to the

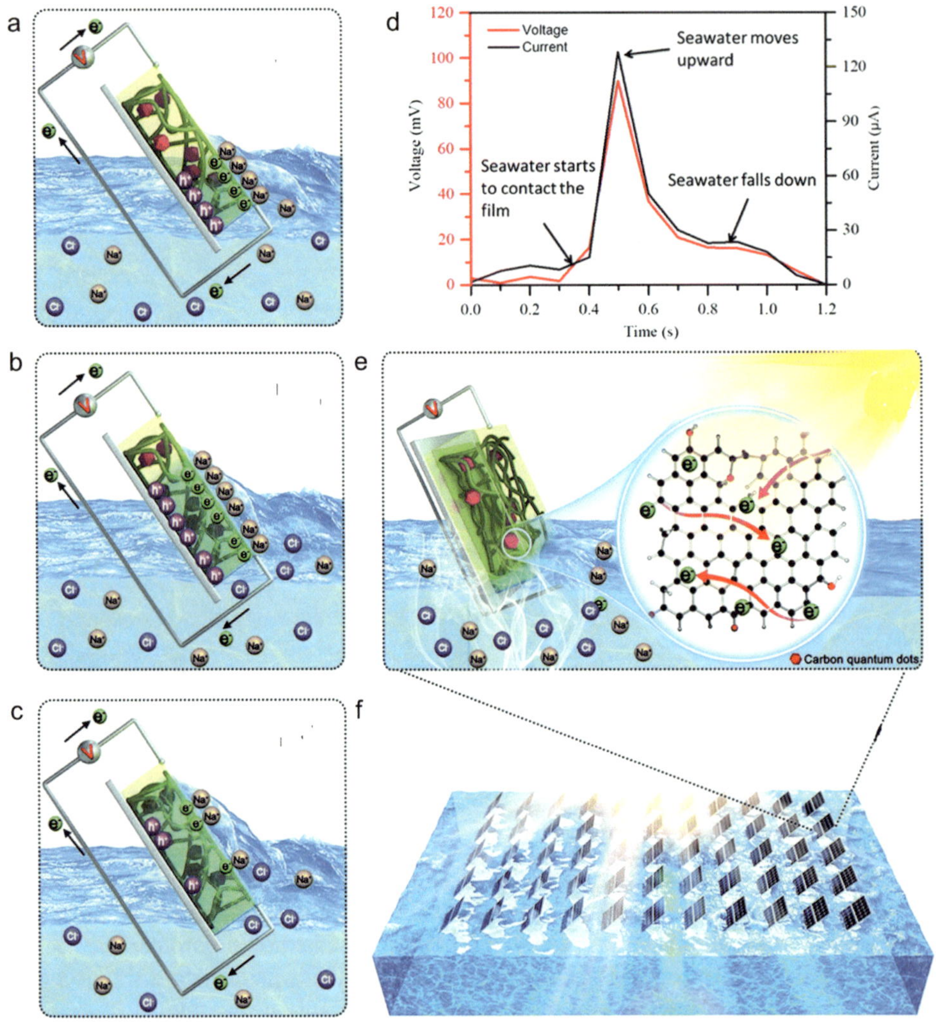

Figure 11.
(a) When the ocean wave meets with the PU/GO-CB-MWCNT/CQDs film, an EDL is generated due to the formation of Na^+ cation layer and electron layer [25]. (b) When the ocean wave reaches the top of the film, a highest voltage can be observed due to the charge of EDL [25]. (c) When the ocean wave is falling downward, a decreasing voltage is obtained because of the discharging of EDL [25]. (d) the voltage and current change of HV cell in one cycle of ocean wave change [25]. (e) under illumination, the electron density is enhanced with CQDs through light excitation [25]. (f) Illustration of the HV networks assembled with HV cells in series and/or parallel. This network can float on the ocean and harvest wave energy and solar energy [25].

highest position, the maximum peak electricity signal is obtained. In this work, CQDs are used for visible light absorption in the range of 330–490 nm. Then more electrons can be excited, the surface electron density increased as in **Figure 11e**. The authors lastly proposed a circuit design for scaling up the power output in large-scale networks (**Figure 11f**). However, a cost-effective, stable, and promising scalable approach for efficiently harvesting ocean wave energy is an open question.

5. Challenges and perspectives

As shown in **Table 1**, the performance of reported HV devices based on the carbon materials is summarized. Even though the great progress has been achieved in recent years, there are several challenges to overcome in the future. The challenges include the following: (i) the power is far from the need of practical applications. (ii) The stability and durability in real environment is not clear. (iii) It is difficult to achieve large-scale integrated applications. (iv) More advanced experimental technologies should be developed to reveal the unclear mechanism. (v) Non-carbon hydrovoltaic materials should be synthesized or constructed.

Nowadays, the research of hydrovoltaic materials and technologies is still in its infancy. To solve the above issues, we may try to follow the following approaches. (i) Understand the interaction mechanism between water and carbon for the further improvement of hydrovoltaic device performance. This requires to controllably modify the electronic structure of carbon and manipulate the molecules/ions in flows. For example, heteroatom doping can be adopted to tune the electric properties. (ii) The composition and nanostructure of carbon materials can significantly affect the capability of electricity generation. Therefore the nanostructure should be optimized to enhance the effective surface area. Moreover, the composition tailoring can enhance the conductivity, reducing the loss during charge transport. (iii) To improve the output of hydrovoltaic devices, the carbon materials can be combined with other functional materials, such as photovoltaic materials, ferroelectric material, and piezoelectric materials. This route can additionally convert the solar energy and mechanical energy for higher voltage and current outputs. (iv) Develop new nanomaterials with hydrovoltaic properties. For example, graphdiyne as a new 2D carbon material has unique sp. and sp^2 hybridized electronic structure, high theoretical conductivity, good chemical activity, good physical stability, and the intrinsic bandgap of ~0.5 eV. Moreover, the synthesis temperature is usually below 100°C. We believe this new carbon material will exhibit unique hydrovoltaic properties. (v) Some in situ and in operando technologies should be used to characterize the interface between water and solid at atom/nano level, such as atomic force microscopy, infrared/Raman spectroscopy, AC impedance spectroscopy, and so on. Nevertheless, hydrovoltaic materials and technologies are very promising for harvesting energy in water. More research efforts should be devoted to realize the practical applications in the near future. In hydrovoltaic field, China stands in the forefront of the world. To realize the practical applications, the multidisciplinary collaboration in research, the government support, and the market promotion are urgent needed in the following decade years.

6. Conclusions

In summary, we introduced the water-carbon interactions and the popular mechanisms of hydrovoltaic effects and reviewed the recent progress of hydrovoltaic devices. This field is in its infancy but is a promising direction in the future. Great achievements in moisture, droplet, flow, and evaporation-induced electricity generation have been gained. Since water evaporation is uninterrupted and available under any conditions, the hydrovoltaic devices have great advantages over other energy conversion devices if the power could meet the daily usage. As in **Table 1**, it is exciting that the optimized hydrovoltaic devices now can generate voltage of 3 V. We believe that the rapid growth will bring this emerging hydrovoltaic device into a viable and broad industry technology.

In China, the water resource is around 6% of the Earth's water resource. However, 80% of the water resource is distributed in the South of China. Therefore the energy harvested from water through hydrovoltaic effect, not the conventional hydropower station, is an attracting and alternative approach in China because the hydrovoltaic effect can generate electricity from not only the water flow, but also the water moisture, droplet, and evaporation. This future technology is a very

Materials	Substrate	Solution	Flow type	Potential (mV)	Current (μA)	Refs.
GO film	/	DI water	Moisture	700	25	[42]
GO framework	/	DI water	Moisture	260	/	[16]
GO film	/	DI water	Moisture	1500	136	[43]
GO film	/	DI water	Moisture	700	0.2	[44]
GO	/	DI water	Moisture	340	~1	[45]
GO nanoribbon	/	DI water	Moisture	40	300	[46]
GQDs	PET	DI water	Moisture	270	/	[14]
Wrinkled graphene	SiO_2/Si	NaCl	Moisture	20	0.045	[17]
Monolayer reduced GO	ITO/PET	Simulated raindrops	Droplet	0.1	0.49	[38]
Graphene/carbon black/PTFE	ITO/FTO	Simulated raindrops	Droplet	0.228	5.97	[18]
Monolayer graphene	SiO_2/Si	DI water	Droplet	28.14	1800	[39]
Monolayer graphene	PVDF	DI water	Droplet	100	/	[47]
Monolayer graphene	SiO_2/Si	DI water	Droplet	10	0.5	[48]
Nitrogen-doped graphene	SiO_2/Si	DI water	Droplet	380	/	[49]
Graphene grid	PDMS	NaCl	Droplet	0.1	/	[50]
Monolayer graphene	SiO_2/Si	NaCl	Droplet	0.15	/	[11]
Monolayer graphene	PET etc.	NaCl	Droplet	500	/	[51]
Graphene foam	/	DI water	Flow	0.001	100	[52]
Reduced GO	Paper-pencil	DI water/$MgCl_2$	Flow	280	812.5	[53]
Graphene/carbon black	Glass, etc.	Simulated waving	Flow	11.14	3	[25]
Monolayer graphene	PET	NaCl	Flow	100	11	[10]
Carbon black	Quartz	DI water	Flow	1000	0.15	[29]
Graphene hydrogel membrane	/	NaCl	Flow	/	0.002	[54]
Few-layer graphene	SiO_2/Si	HCl	Flow	25	340	[55]
Few-layer graphene	SiO_2/Si	HCl	Flow	120	/	[56]
Pair of graphene sheets	/	NaCl	Flow	1000	2	[24]
Aligned CNT fiber	/	NaCl	Flow	341	/	[22]
Carbon black film		DI water	Evaporation	1000	/	[29]
Carbon Film	Al_2O_3	DI water	Evaporation	1000	0.6	[57]
Graphene/carbon cloth	/	NaCl	Evaporation	370	/	[58]
Carbon black-glass fiber hybrid film	/	DI water	Evaporation	3000	/	[30]
Partially reduced GO sponge	/	DI water	Evaporation	630	~100	[59]

PMMA, polymethyl methacrylate; PVDF, polyvinylidene fluoride; ITO, indium tin oxide; FTO, fluorine-doped tin oxide; PDMS, polydimethylsiloxane.

Table 1.
Performance summary of reported hydrovoltaic devices based on the carbon materials.

promising solution in the North of China. More importantly, this future technology can harvest ocean wave energy for island power supply in Chinese waters.

Acknowledgements

The authors are grateful for the financial support from the National Natural Science Foundation of China (Grant 21703150), the China Postdoctoral Science Foundation (Grant 2015 M582495), and the Sichuan Science and Technology Program (Grant 2018JY0015).

Conflict of interest

The authors declare no conflict of interest.

Author details

Jiale Xie*, Liuliu Wang, Xiaoying Chen, Pingping Yang, Fengkai Wu and Yuelong Huang
Institute of Photovoltaics, Southwest Petroleum University, Chengdu, People's Republic of China

*Address all correspondence to: jialexie@swpu.edu.cn

References

[1] Zhang Z et al. Emerging hydrovoltaic technology. Nature Nanotechnology. 2018;**13**(12):1109-1119. DOI: 10.1038/s41565-018-0228-6

[2] Roger I, Shipman MA, Symes MD. Earth-abundant catalysts for electrochemical and photoelectrochemical water splitting. Nature Reviews Chemistry. 2017;**1**(1):0003. DOI: 10.1038/s41570-016-0003

[3] Salamanca JM, Alvarez-Silva O, Tadeo F. Potential and analysis of an osmotic power plant in the Magdalena River using experimental field-data. Energy. 2019;**180**:548-555. DOI: 10.1016/j.energy.2019.05.048

[4] Husain AAF et al. A review of transparent solar photovoltaic technologies. Renewable and Sustainable Energy Reviews. 2018;**94**:779-791. DOI: 10.1016/j.rser.2018.06.031

[5] Han Y, Zhang Z, Qu L. Power generation from graphene-water interactions. FlatChem. 2019;**14**:100090. DOI: 10.1016/j.flatc.2019.100090

[6] Ghosh S, Sood AK, Kumar N. Carbon nanotube flow sensors. Science. 2003;**299**(5609):1042-1044. DOI: 10.1126/science.1079080

[7] Xu Y, Chen P, Peng H. Generating electricity from water through carbon nanomaterials. Chemistry – A European Journal. 2018;**24**(24):6287-6294. DOI: 10.1002/chem.201704638

[8] Tang Q, Yang P. The era of water-enabled electricity generation from graphene. Journal of Materials Chemistry A. 2016;**4**(25):9730-9738. DOI: 10.1039/C6TA03107B

[9] Stone HA, Stroock AD, Ajdari A. Engineering flows in small devices: Microfluidics toward a lab-on-a-Chip. Annual Review of Fluid Mechanics. 2004;**36**(1):381-411. DOI: 10.1146/annurev.fluid.36.050802.122124

[10] Yin J et al. Waving potential in graphene. Nature Communications. 2014;**5**(1):3582. DOI: 10.1038/ncomms4582

[11] Yin J et al. Generating electricity by moving a droplet of ionic liquid along graphene. Nature Nanotechnology. 2014;**9**:378. DOI: 10.1038/nnano.2014.56

[12] Zhao F et al. Direct power generation from a graphene oxide film under moisture. Advanced Materials. 2015;**27**(29):4351-4357. DOI: 10.1002/adma.201501867

[13] Ponomarenko LA et al. Chaotic Dirac Billiard in graphene quantum dots. Science. 2008;**320**(5874):356-358. DOI: 10.1126/science.1154663

[14] Huang Y et al. Highly efficient moisture-triggered nanogenerator based on graphene quantum dots. ACS Applied Materials & Interfaces. 2017;**9**(44):38170-38175. DOI: 10.1021/acsami.7b12542

[15] Liu K et al. Induced potential in porous carbon films through water vapor absorption. Angewandte Chemie International Edition. 2016;**55**(28):8003-8007. DOI: 10.1002/anie.201602708

[16] Zhao F et al. Highly efficient moisture-enabled electricity generation from graphene oxide frameworks. Energy & Environmental Science. 2016;**9**(3):912-916. DOI: 10.1039/C5EE03701H

[17] Zhen Z et al. A non-covalent cation-π interaction-based humidity-driven electric nanogenerator prepared with salt decorated wrinkled graphene.

Nano Energy. 2019;**62**:189-196. DOI: 10.1016/j.nanoen.2019.05.026

[18] Tang Q et al. An all-weather solar cell that can harvest energy from sunlight and rain. Nano Energy. 2016;**30**:818-824. DOI: 10.1016/j.nanoen.2016.09.014

[19] Wang Y et al. Harvest rain energy by polyaniline-graphene composite films. Renewable Energy. 2018;**125**:995-1002. DOI: 10.1016/j.renene.2018.03.034

[20] Li J et al. Electricity generation from water droplets via capillary infiltrating. Nano Energy. 2018;**48**:211-216. DOI: 10.1016/j.nanoen.2018.02.061

[21] Tang W, Chen BD, Wang ZL. Recent progress in power generation from water/liquid droplet interaction with solid surfaces. Advanced Functional Materials. 2019;**29**(41):1901069. DOI: 10.1002/adfm.201901069

[22] Xu Y et al. A one-dimensional fluidic nanogenerator with a high power conversion efficiency. Angewandte Chemie International Edition. 2017;**56**(42):12940-12945. DOI: 10.1002/anie.201706620

[23] Liu J, Dai L, Baur JW. Multiwalled carbon nanotubes for flow-induced voltage generation. Journal of Applied Physics. 2007;**101**(6):064312. DOI: 10.1063/1.2710776

[24] Fei W et al. Waving potential at volt level by a pair of graphene sheets. Nano Energy. 2019;**60**:656-660. DOI: 10.1016/j.nanoen.2019.04.020

[25] Tan J et al. Generators to harvest ocean wave energy through electrokinetic principle. Nano Energy. 2018;**48**:128-133. DOI: 10.1016/j.nanoen.2018.03.032

[26] He S et al. Chemical-to-electricity carbon: Water device. Advanced Materials. 2018;**30**(18):1707635. DOI: 10.1002/adma.201707635

[27] Pei J et al. Liquid flow-induced electricity in carbon nanomaterials. Sustainable Energy & Fuels. 2019;**3**(3):599-610. DOI: 10.1039/C8SE00604K

[28] Tarelho JPG et al. Graphene-based materials and structures for energy harvesting with fluids – A review. Materials Today. 2018;**21**(10):1019-1041. DOI: 10.1016/j.mattod.2018.06.004

[29] Xue G et al. Water-evaporation-induced electricity with nanostructured carbon materials. Nature Nanotechnology. 2017;**12**:317. DOI: 10.1038/nnano.2016.300

[30] Li J et al. Surface functional modification boosts the output of an evaporation-driven water flow nanogenerator. Nano Energy. 2019;**58**:797-802. DOI: 10.1016/j.nanoen.2019.02.011

[31] Xue J et al. Vapor-activated power generation on conductive polymer. Advanced Functional Materials. 2016;**26**(47):8784-8792. DOI: 10.1002/adfm.201604188

[32] Yavari F et al. Tunable bandgap in graphene by the controlled adsorption of water molecules. Small. 2010;**6**(22):2535-2538. DOI: 10.1002/smll.201001384

[33] Bi H et al. Ultrahigh humidity sensitivity of graphene oxide. Scientific Reports. 2013;**3**:2714. DOI: 10.1038/srep02714

[34] Borini S et al. Ultrafast graphene oxide humidity sensors. ACS Nano. 2013;**7**(12):11166-11173. DOI: 10.1021/nn404889b

[35] Chen X et al. Scaling up nanoscale water-driven energy conversion into evaporation-driven engines and generators. Nature Communications. 2015;**6**(1):7346. DOI: 10.1038/ncomms8346

[36] Cheng H et al. Graphene fibers with predetermined deformation as moisture-triggered actuators and robots. Angewandte Chemie International Edition. 2013;**52**(40):10482-10486. DOI: 10.1002/anie.201304358

[37] Liu Z et al. Surface-energy generator of single-walled carbon nanotubes and usage in a self-powered system. Advanced Materials. 2010;**22**(9):999-1003. DOI: 10.1002/adma.200902153

[38] Tang Q et al. A solar cell that is triggered by sun and rain. Angewandte Chemie International Edition. 2016;**55**(17):5243-5246. DOI: 10.1002/anie.201602114

[39] Zhong H et al. Graphene based two dimensional hybrid nanogenerator for concurrently harvesting energy from sunlight and water flow. Carbon. 2016;**105**:199-204. DOI: 10.1016/j.carbon.2016.04.030

[40] Wang ZL. Catch wave power in floating nets. Nature. 2017;**542**(7640):159-160. DOI: 10.1038/542159a

[41] Tan J et al. Photo-induced charge boosting of liquid–solid electrokinetic generators for efficient wave energy harvesting. Journal of Materials Chemistry A. 2019;**7**(10):5373-5380. DOI: 10.1039/C8TA12037D

[42] Xu T et al. Electric power generation through the direct interaction of pristine graphene-oxide with water molecules. Small. 2018;**14**(14):1704473. DOI: 10.1002/smll.201704473

[43] Huang Y et al. Interface-mediated hygroelectric generator with an output voltage approaching 1. 5 volts. Nature Communications. 2018;**9**(1):4166. DOI: 10.1038/s41467-018-06633-z

[44] Liang Y et al. Electric power generation via asymmetric moisturizing of graphene oxide for flexible, printable and portable electronics. Energy & Environmental Science. 2018;**11**(7):1730-1735. DOI: 10.1039/C8EE00671G

[45] Shao C et al. Wearable fiberform hygroelectric generator. Nano Energy. 2018;**53**:698-705. DOI: 10.1016/j.nanoen.2018.09.043

[46] Zhao F et al. Graphene oxide nanoribbon assembly toward moisture-powered information storage. Advanced Materials. 2017;**29**(3):1604972. DOI: 10.1002/adma.201604972

[47] Zhong H et al. Graphene-piezoelectric material heterostructure for harvesting energy from water flow. Advanced Functional Materials. 2017;**27**(5):1604226. DOI: 10.1002/adfm.201604226

[48] Zhong H et al. Two dimensional graphene nanogenerator by coulomb dragging: Moving van der Waals heterostructure. Applied Physics Letters. 2015;**106**(24):243903. DOI: 10.1063/1.4922800

[49] Okada T et al. Role of doped nitrogen in graphene for flow-induced power generation. Advanced Engineering Materials. 2018;**20**(11):1800387. DOI: 10.1002/adem.201800387

[50] He Y et al. Galvanism of continuous ionic liquid flow over graphene grids. Applied Physics Letters. 2015;**107**(8):081605. DOI: 10.1063/1.4929745

[51] Yang S et al. Mechanism of electric power generation from ionic droplet motion on polymer supported graphene. Journal of the American Chemical Society. 2018;**140**(42):13746-13752. DOI: 10.1021/jacs.8b07778

[52] Huang W et al. Power generation from water flowing through three-dimensional graphene foam. Nanoscale. 2014;**6**(8):3921-3924. DOI: 10.1039/C3NR04261H

[53] Arun RK et al. Energy generation from water flow over a reduced graphene oxide surface in a paper–pencil device. Lab on a Chip. 2016;**16**(18): 3589-3596. DOI: 10.1039/C6LC00820H

[54] Guo W et al. Bio-inspired two-dimensional nanofluidic generators based on a layered graphene hydrogel membrane. Advanced Materials. 2013;**25**(42):6064-6068. DOI: 10.1002/adma.201302441

[55] Dhiman P et al. Harvesting energy from water flow over graphene. Nano Letters. 2011;**11**(8):3123-3127. DOI: 10.1021/nl2011559

[56] Yin J et al. Harvesting energy from water flow over graphene? Nano Letters. 2012;**12**(3):1736-1741. DOI: 10.1021/nl300636g

[57] Ding T et al. All-printed porous carbon film for electricity generation from evaporation-driven water flow. Advanced Functional Materials. 2017;**27**(22):1700551. DOI: 10.1002/adfm.201700551

[58] Zhang L, Chen X. Nanofluidics for giant power harvesting. Angewandte Chemie International Edition. 2013;**52**(30):7640-7641. DOI: 10.1002/anie.201302707

[59] Zhang G et al. Harvesting environment energy from water-evaporation over free-standing graphene oxide sponges. Carbon. 2019;**148**:1-8. DOI: 10.1016/j.carbon.2019.03.041

Chapter 4

A Circular Economy of Electrochemical Energy Storage Systems: Critical Review of SOH/RUL Estimation Methods for Second-Life Batteries

Simon Montoya-Bedoya, Laura A. Sabogal-Moncada, Esteban Garcia-Tamayo and Hader V. Martínez-Tejada

Abstract

Humanity is facing a gloomy scenario due to global warming, which is increasing at unprecedented rates. Energy generation with renewable sources and electric mobility (EM) are considered two of the main strategies to cut down emissions of greenhouse gasses. These paradigm shifts will only be possible with efficient energy storage systems such as Li-ion batteries (LIBs). However, among other factors, some raw materials used on LIB production, such as cobalt and lithium, have geopolitical and environmental issues. Thus, in a context of a circular economy, the reuse of LIBs from EM for other applications (i.e., second-life batteries, SLBs) could be a way to overcome this problem, considering that they reach their end of life (EoL) when they get to a state of health (SOH) of 70–80% and still have energy storage capabilities that could last several years. The aim of this chapter is to make a review of the estimation methods employed in the diagnosis of LIB, such as SOH and remaining useful life (RUL). The correct characterization of these variables is crucial for the reassembly of SLBs and to extend the LIBs operational lifetime.

Keywords: second-life batteries, RUL/SOH estimation, circular economy, energy storage, Li-ion battery

1. Introduction

The Sustainable Development Goals (SDGs) are a call to action against global issues in the twenty-first century [1] such as climate change, geopolitical topics, overgrowing population, increasing energy demand, and resource scarcity, among others [2]. According to the International Energy Agency (IEA) statistics, the electricity and heat producers and transport sector are the largest greenhouse gas emitters, with at least 90% of the total CO_2 emissions [3, 4]. In 2018, above 26% of electric energy was generated from renewable sources (RSs) [5]. However, this percentage is still low in order to maintain global warming below the 2°C increment threshold stated in the 2015 Paris Agreement [6]. Taking this into account, its clear that humanity must implement disruptive strategies to tackle these challenges. In this regard, electricity generation with RSs and electric mobility (EM) have become two of the main mechanisms for the decarbonization of the power and mobility sectors. In this context, electrochemical energy storage (EES) is a fundamental technology to realize these energy transitions by coupling both sectors in this time in history and transforming RSs from an alternative to a reliable source.

The most familiar EES devices are batteries. Compared to other energy storage mechanisms, the energy capacity of batteries is relatively low, but its efficiency is high (>95%) [7]. This makes batteries an ideal energy storage system for small- and large-scale applications [8]. According to Garcia-Tamayo [9], the convenience of batteries lies in the wide range of sizes in which they may be manufactured or assembled into packs, their ability to supply electrical power instantly, their portability (for smaller sizes), and the option of single-use or multiple-use units. The World Economic Forum reported that batteries could enable 30% of the required CO_2 reductions in the transport and power sectors, provide access to electricity for 600 million people who currently lacking access, and create 10 million safe and sustainable jobs around the world [10]. Also, since the use of internal combustion engine (ICE) vehicles accounts for a large portion of the daily energy consumption, a continuous increase of batteries through electric vehicle (EV) adoption might lead to improve grid stabilization.

Li-ion batteries (LIBs) are the most common batteries available at present and are found in almost all commercial EVs today. The battery packs inside a vehicle are composed of modules connected in series or parallel to reach the energy output and power required. Each module, on its turn, is also composed of Li-ion cells connected in series or parallel. Thus, a Li-ion cell acts as a fundamental brick of today's battery systems. A schematic illustration can be found in **Figure 1**. When an electrical load (i.e., electric vehicle, solar panel/electrical grid) is plugged and the circuit closed, during discharge, electrons (green circles) flow from the anode to cathode creating an electronic current. Likewise, Li-ions (yellow circles) are flowing in the same direction (from anode to cathode), thus converting chemical energy into electrical energy. Ions move between the electrodes by means of an electrolyte which has the property to be a good electronic insulator and good ionic conductor. As a liquid electrolyte is used in most of the cases, a separator is placed in the middle in order to maintain an even spacing between both electrodes. This separator must provide blocking of electronic current and permeation of its ionic analogue. The process shown in the schematic occurs during cell discharge. During charge, an external voltage is applied to the circuit, forcing electrons and

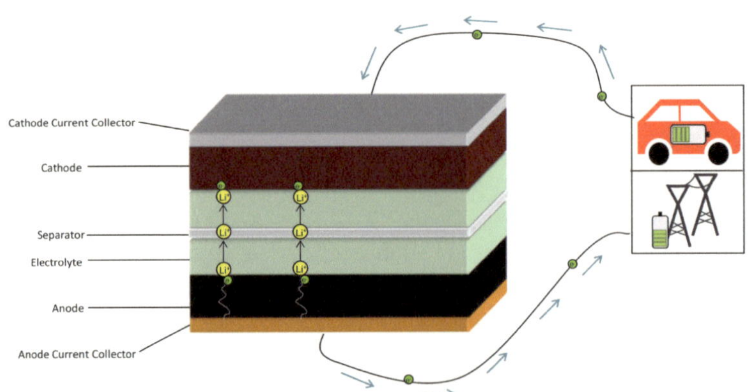

Figure 1.
Schematic of a Li-ion cell during discharge.

ions to flow from the cathode to anode. This process is performed to convert electrical energy back to chemical energy.

In general, commercial LIBs have highly pure graphite as active material for anode and different transition metal oxide lithium compounds as active material for cathode, such as $LiNi_{0.33}Mn_{0.33}Co_{0.33}O_2$ (NMC 111), $LiFePO_4$ (LFP), and $LiCoO_2$ (LCO), among others. All these cathode materials are found in commercial batteries and are referred to in the literature as battery or cathode chemistries. However, it is important to clarify that all of them are LIB technologies.

Despite the positive attributes previously described for LIB systems, there are also a set of critical characteristics that affect the battery behavior with time and as a result of their usage. The sum of these effects is a process commonly referred in literature as battery degradation or aging, which affects the cells' ability to store energy and meet power demands and, ultimately, leads to their end of life (EoL). LIBs are sensitive to the way they are charged and discharged, especially in extreme conditions such as overcharge and deep discharge as they increase the aging effect. Thus, it is of outmost importance for any device powered by LIBs to be informed of the amount of energy that can be stored and the power that can be provided by the battery at any point in time. However, the rates at which these variables degrade over time cannot be directly measured in real applications, so they must be inferred indirectly using methods and models that use input data that can be measured during charge or discharge operations.

Degradation in Li-ion cells is caused by a large number of physical and chemical mechanisms, such as active material particle cracking during Li-ion insertion and de-insertion, formation of a passivating layer on the anode/electrolyte interphase during the first cycles (solid electrolyte interphase, SEI), SEI decomposition and precipitation in the electrolyte, lithium plating and dendrite formation that could cause internal short circuit, and dissolution of transition metals from the cathode in the electrolyte, among multiple others. Multiple reviews can be found in the literature summarizing and describing in detail these aging mechanisms [11–13].

Fabrication of LIBs uses key and critical raw materials, whose exploitation and market are associated to unequal distribution of the mineral resources in the world [14]. Although lithium is a key ingredient in LIBs, manufacturers commonly use lithium carbonate or lithium hydroxide in batteries rather than pure metallic lithium. They also include other metals, such as cobalt, graphite, manganese, and nickel. Among them, cobalt and lithium are the most constrained materials [15], and nickel is important in recycling and is highly toxic to the environment. According to the US Geological Survey, worldwide lithium supply had an increase of around 23% from 2017 to 2018, coming in at 85,000 metric tons (MT) of lithium content [16]. Harper et al. estimated that the 1 million EVs that were sold in 2017 together account for nearly 250,000 MT of batteries [17]. BloombergNEF recently reported that 2 million EVs were sold in 2018, from just a few thousand in 2010, and there is no sign of slowing down. Annual passenger EV sales are forecasted to rise to 10 million in 2025, 28 million in 2030, and 56 million by 2040 [18].

In a rough approximation, if a full electric vehicle with a 33-kWh battery pack requires ≈ 5.3 kg of Li, just the EVs sold in 2018 may have required ≈10,600 MT of lithium content. If battery capacities will have an increase of at least 1.8 times by 2025 (i.e., in 6 years the capacity for the Ford Focus EV raised from 23 kWh in 2012 to 33.5 kWh in 2018, while the Renault Zoe changed from 22 kWh in 2012 to 51 kWh in 2019), the EV market will require ≈93,000 MT of lithium content (assuming the design or battery chemistries will not change over time), exceeding current world production. Therefore, there is still not a clear way to use less metal without compromising life span or energy storage capacity.

At present, EV batteries, most of them based on Li-ion technology, have a useful lifetime (defined by the loss of capacity due to degradation until they reach 80% of their nominal capacity) of around 300–15,000 cycles, depending on the conditions in which the battery is charged and discharged [19]. However, it is likely that they will be changed before they reach the 80% threshold not because they do not work properly but because there are other battery technologies and chemistries that will get better in the near future. For example, a recent study by Professor Jeff Dahn's group at Dalhousie University and Tesla Canada presented a LIB testing benchmark where they included a battery with a lifetime of around 4600 cycles (1.600,000 km driving range), at extreme discharging conditions (i.e., bringing the battery to a full discharge in each cycle), which could also be employed in energy storage for 20 years after reaching its EV end of life [20]. Still, even if novel batteries will get more cost-effective and safer, the battery manufacturing processes remain energy-intensive [21].

When EV batteries reach their end of life, i.e., when they reach the 80% threshold, they can still store enough energy and can operate perfectly in other uses, opening the possibility to extend their operational lifetime into a second one. Such use has been recently termed as second-life batteries (SLBs). SLB management and their possible applications are receiving a lot of attention because they could serve as a tool against the issue of 'waste' batteries being stored before repurposing or final disposal and could also save many tasks related with the managing, chemical and mechanical dismantling, and separation processes that recycling entails. To put it in perspective, the future 10 million EVs that will be sold in 2025 [18] account for nearly ≈2,200,000 MT of batteries [22], which, in the absence of a second life, would otherwise end up as waste. Moreover, in the waste management hierarchy, reuse is considered preferable to recycling [17].

According to the Advanced Battery Consortium (USABC), and in most literature related to electric mobility [23], the end of life for an EV battery is defined as a 20% drop of cell capacity from the nominal value or a 20% drop from the rated power density at 80% depth of discharge (DoD, defined as the fraction or percentage of the capacity which has been removed from the fully charged battery). Nonetheless, among other factors, from an electrical and electrochemical standpoint, in order to classify the delivery of SLBs as a capable and efficient energy storage system, its remaining capacity, power, and functionality must be properly identified.

A circular economy framework diagram for LIBs is shown in **Figure 2**: (i) Used batteries from EVs that have reached their end of first life are collected. Usually their state of health (SOH) is unknown but should be around 80%. (ii) SOH testing of the battery pack/module/cell is needed to characterize its remaining capacity as compared to its initial capacity. (iii.a) The battery is depleted if the SOH is less than 40%, (iii.b) It is still usable if SOH is greater than 40%. (iv) The battery is sent for repurposing. If needed it might be broken down into its fundamental parts (cells) to connect it in series or parallel to obtain the desired energy output power for each specific application. (v) At this point, the repurposed system becomes a second-life battery and is placed on the market as a new product to serve in a second-life application. (vi) The SLB is collected after reaching its end of second life, and step (ii) is repeated to check if a third-life application is possible. (vii) If not, the battery is sent for recycling where the raw materials will be recovered and restored. Finally, the recovered materials are sent for the remanufacture of new products such as the production of new Li-ion batteries (where the whole cycle would start over).

It is important to remark that step (ii), i.e., SOH testing, is crucial to determine if the battery is depleted and immediately goes to recycling or if it may be used as a SLB for other applications. In this chapter, we will review the diagnostic and prognostic methods needed to estimate the battery current storage capacity, the

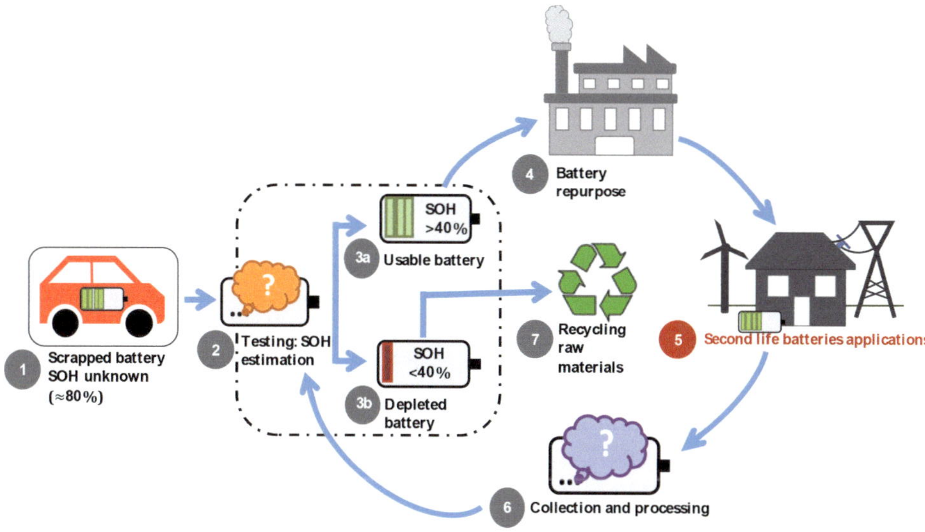

Figure 2.
Li-ion battery circular economy framework diagram.

state of health, and the remaining useful life (RUL), which are key variables that will provide the inputs needed to define possibilities for SLB applications.

2. Review methodology

A systematic review methodology was employed as a screening method to select the information. Scopus was used as scientific database, using the following keywords as query string: Li-ion-batteries AND soh OR rul AND estimation methods AND electrochemical model OR second-life batteries. These keywords were chosen to narrow the scope of this review chapter to those focusing only on estimation methods that could be extended from SOH percentages below the 70–80% electric mobility threshold to scenarios for stationary energy storage applications that use SOH percentages that can go as low as 40%. This screening method resulted in 152 articles. A further selection was done after analyzing the title, abstract, keywords, and paper content. We identified and analyzed 15 papers which included journals and conference proceedings. The selected 15 references were studied in detail to extract useful information such as type of estimation method, estimated variables (SOH/RUL), experimental conditions, minimum SOH reached, and reported error.

3. Estimation methods

Before reviewing and establishing a classification of the estimation methods, it is important to provide definitions of the main variables found in the literature.

State of health is a percentage that measures the remaining capacity of an aged battery compared to the capacity when it was fresh. It is defined by Eq. (1).

$$SOH = \frac{Q_{actual}}{Q_{nominal}} \times 100\% \tag{1}$$

where Q_{actual} and $Q_{nominal}$ represent the actual capacity and the nominal capacity, respectively.

Remaining useful life is an estimation of the remaining time or number of cycles until the SOH of a battery reaches a specific threshold usually defined by an application. For example, in electric mobility, it is calculated until the SOH reaches 80%. Although in the literature some authors define the RUL as the time in which the SOH of the batteries reaches 0% [24], there are few articles in which the SOH is estimated below the 80% threshold.

One of the main aspects for RUL estimation is to have an accurate knowledge of the current battery state of health [25]. In the case of RUL for SLBs, it is crucial to know the minimum SOH requirements for each application in order to estimate the number of cycles or the remaining time that the batteries will last [26, 27].

In general, estimation methods for SOH and RUL are described separately in the literature [28–30]. Some authors have classified battery models for SOH diagnosis as *electrochemical*, *electrical*, and *mathematical models* [31], while others have grouped them as *direct measurements*, *model-based*, and *adaptive techniques* [32]. Similar categorizations can be found in the literature for RUL estimation methods and have been organized as *adaptive filter*, *intelligent*, and *stochastic techniques* [28]. Particularly, the classifications made by Saidani et al. [33] and Liao et al. [34] are interesting as they introduce a comprehensible way to group both SOH and RUL estimation methods in three categories, based on system theory concepts: *white-box*, *black-box*, and *gray-box* methods (see **Figure 3**). In general, these concepts refer to the level of theoretical or experimental knowledge needed to describe or model a process. Each set will be discussed in detail, but in summary white-box methods try to elucidate what happens inside a battery in terms of aging and degradation, while black-box methods employ mathematical and stochastic equations to establish correlations between intrinsic electrochemical mechanisms and external variables that can be easily measured. Gray-box methods are hybrid prognostics between white- and black-box methods where both internal mechanisms of batteries and data-driven models are integrated.

3.1 White-box methods

White-box models refer to methods that consider internal reactions and aging mechanisms of the batteries, which include physicochemical, electrochemical, and thermodynamic theories [35]. For instance, Fu et al. [36] developed a degradation model based on partial differential equations (PDEs) that estimate the capacity

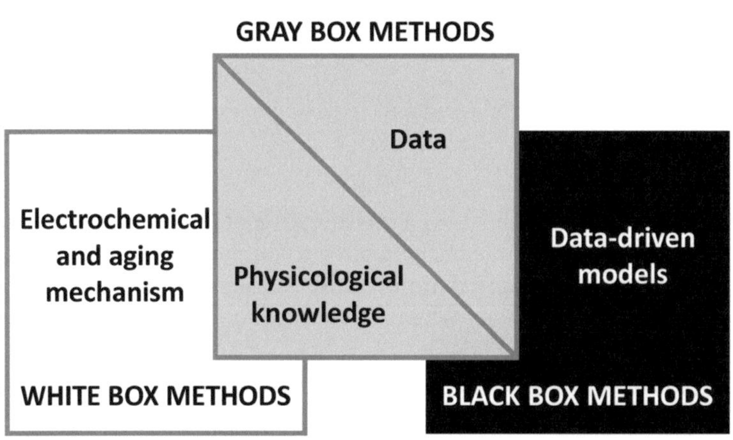

Figure 3.
Classification of SOH and RUL estimation methods.

fade using three key parameters: (i) the volume fraction of accessible material in the anode, (ii) ionic and electronic resistance of the solid electrolyte interphase and deposited layers on the electrode surfaces, and (iii) diffusion coefficient of the electrolyte. These parameters must be estimated through experimental tests and validated by characterization techniques such as scanning electron microscopy, X-ray diffraction, or X-ray photoelectron spectroscopy for each battery chemistry. This model exemplifies two of the main disadvantages of white-box methods: the need to estimate a lot of parameters and the solution of complex PDE systems. Most of the times, white-box methods derive results that are not cost-effective [33, 37].

Similarly, Gao et al. [38] proposed an electrochemical aging model that estimates the capacity fade considering the change of the open-circuit voltage (OCV) over the life span of a Li-ion battery. They reported a maximum error of about 2% for different batteries charged and discharged at different current rates (C-rates), namely, 1C, 2C, and 3C. However, this error tends to increase at the final phase of the cycling test. Likewise, with the purpose of reducing the complexity of electrochemical models, there are other methods such as single-particle models (SPMs), which assume each electrode as a single particle in order to obtain an ordinary differential equation system that models the Li-ion battery behavior [39–41]. SPMs have been integrated with a capacity degradation model coupled to a chemical/mechanical degradation mechanism that allows the prediction of the capacity fade as a function of battery temperature and cycling. The root mean squared errors (RMSEs) in these estimation methods were $\mathbf{7.21 \times 10^{-3}}$, $\mathbf{7.43 \times 10^{-3}}$, and $\mathbf{10.3 \times 10^{-3}}$ for $LiFePO_4$ (cathode)/graphite (anode) batteries tested at 15, 45, and 60°C, respectively [42].

On the other hand, white-box methods have not been used for RUL estimation due to the reasons mentioned above, i.e., because of the complexity of the models and the fact that cycles are not explicit on most of this type of methods. Thus, it is difficult to obtain parameters for SLBs' RUL because the information of the batteries on their fresh state is normally unknown [43]. However, some authors have used empirical approximations, such as Arrhenius equation (takes temperature as an accelerated aging factor) and power law (takes mechanical/electrical stress as an accelerated aging factor), to model capacity loss on batteries as a function of cycle number [30, 44].

As a result, the implementation of these methods on SLBs has been relegated since most of them do not consider the C-rate as an explicit parameter on their aging models. SOH and RUL estimation for SLBs should consider the load profile of each future application in terms of the current (amperes) needed [26, 45]. These methods have been developed for automotive applications where batteries reach their EoL when they get to a state of health of 70–80% [46] and where the capacity degradation is approximately linear until this SOH threshold, as shown in **Figure 4**. After this point, the aging behavior changes and nonlinearities start to appear [47, 48].

3.2 Black-box methods

Black-box methods take advantage of data-driven models that establish relationships between unknown intrinsic electrochemical mechanisms and external measurable variables of a Li-ion battery (e.g., voltage, current, temperature, capacity) [23]. These methods extract relevant aging features and construct degradation models based on mathematical and stochastic equations to estimate the SOH and thus predict the RUL [49]. Indeed, aging feature extraction is crucial to obtain

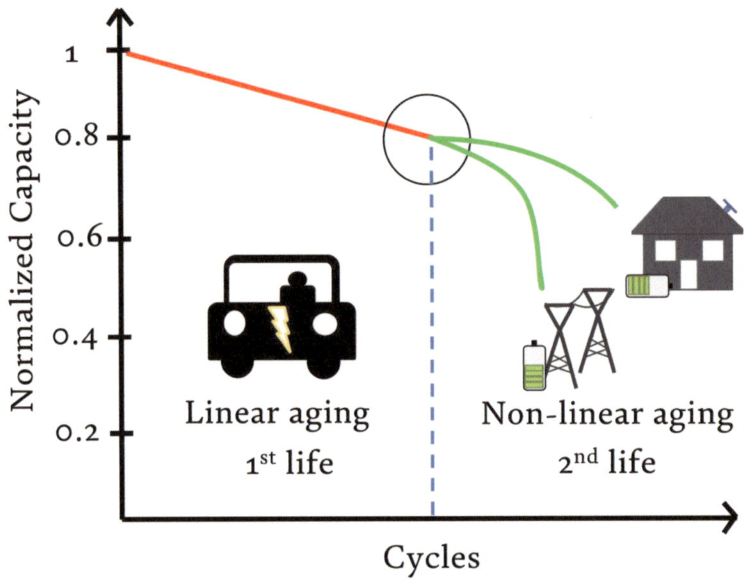

Figure 4.
Illustrative capacity degradation curve for a common Li-ion battery at its first and second life.

accuracy estimations with these kinds of methods [50]. Jiang et al. [51] tested six $LiFePO_4$ batteries, scrapped from a retired battery pack of an EV, with different load profiles simulating frequency regulation and peak shaving applications. They used the incremental capacity analysis (ICA) obtained from a curve of voltage (V) vs. charge/discharge capacity (Q) using Eq. (2), to develop a linear regression, constructed with the ordinary least squares (OLS) method, that could correlate features from the IC curve and the battery SOH. They obtained a mean absolute error and maximum error within 2%. Similarly, Quinard et al. [52] concluded that the ICA technique, used for SOH estimation in SLBs, has a high dependence on the C-rate (i.e., an inverse relationship between C-rate and accuracy). They reported a maximum absolute error of 5%.

$$IC = \frac{dQ}{dV} \tag{2}$$

Likewise, machine learning algorithms have been widely used in battery prognostics as these techniques can extract patterns from battery datasets, such as those from NASA [53] and University of Oxford [54], where batteries were tested at different aging conditions (C-rates and temperature). Support vector machines (SVM) [55], artificial neural networks [56, 57], and fuzzy logic [58] are some of the strategies used for SOH estimation. Nevertheless, to guarantee low-error predictions and robustness against noise, machine learning algorithms need an amount of cycling data corresponding to at least 25% of the whole battery life span [59], which could take months or years to be generated.

Taking this into account, Cai et al. [60] developed a novel method based on a combination of SVM for regression (SVR) and a genetic algorithm that employs short-term features extracted from the voltage response under a current pulse test that lasts just 18 seconds. Therefore, this process can be implemented in real SLB applications. As a result, they obtained a minor RMSE of $\mathbf{19.12 \times 10^{-3}}$ for a battery with a LiFePO4 chemistry compared to a RMSE of $\mathbf{24.8 \times 10^{-3}}$ obtained by a traditional SVR-based model for a $LiCoO_2$ chemistry [61].

Another strategy that has been used to address the issues for these data-driven methods was proposed by Tang et al. [62]. They developed a model migration-based algorithm to predict the battery aging trajectory and the RUL with a notable reduction of experimental tests. This approach generates a well-known base model with enough data that is then employed in an analogous process with less available data. In this case, the base model takes advantage of accelerated aging tests, while the analogous process uses normal aging tests. As a highlight, they reached a RMSE of about 2% in RUL prediction making use of 15% of the aging data.

It is important to mention that some data-driven models extract multiple features from LIBs that do not necessarily enhance the prediction due to an emergence of redundant information [63], whereby a sliding window-based feature extraction [63] and false nearest neighbor [64] algorithms have been implemented.

3.3 Gray-box methods

Gray-box methods are hybrid prognostics between white and black methods. In other words, this category integrates both internal mechanisms of batteries and data-driven models. Liao et al. [65] stipulated that including general aging progression (white-box methods) improves the prediction accuracy of black-box methods. Equivalent circuit models (ECM) have been commonly used to simulate internal parameters such as electrochemical systems in battery management systems (BMS) [45, 66]. For instance, Wei et al. [61] modeled the capacity and impedance degradation parameters using SVR and ECM, respectively. Also, they employed particle filter (PF) to improve the SVR simulation. Tracking these aging characteristics, they estimated SOH and RUL with a high accuracy compared to an artificial neural network-based model. Likewise, references [67, 68] developed a promising modified PF algorithm that avoids particle degradation. For example, Shi et al. [68] demonstrated that improved unscented PF (IUPF) had better accuracy than unscented Kalman filter (UKF) and unscented particle filter (UPF) model prediction of ohmic internal resistance (R_o) and SOH.

In the same way, Tian et al. [69] tested three commercial $LiNi_{0.33}Mn_{0.33}Co_{0.33}O_2$ (NMC) batteries, considering the effect of temperature and discharge rate on aging cycle, to develop an on-board SOH estimation. Their model consisted in a fractional order model (FOM) using Thevenin ECM with the forgetting factor recursive expanded least square (FFRELS) method to estimate the open-circuit voltage which was then correlated to the SOH using the ICA method. Their proposed method obtained a capacity fade with an error of less than 3.1%, independent of the C-rate aging cycles.

Similarly, Guo et al. [70] used an EDKF-based model and second-order RC circuit model to estimate the SOH, obtaining a maximum error below 4%. Hu et al. [71] achieved accurate results for SOH estimation with a relative error within 3%, using a modified moving horizon estimation (mMHE) method integrated to first-order RC ECM.

3.4 Estimation method summary

A comparative summary of the SOH and RUL estimation methods mentioned above, which are included in the 15 references that resulted from the screening method described in the review methodology, can be seen in **Table 1**. For each method, it compares the employed aging feature and the reported error. Finally, there is a column for the minimum SOH reached in order to identify promising methods for SLB estimation.

Authors	Estimation method	Estimated variables	Experimental conditions*	Aging features employed for estimation	Minimum SOH reached	Reported error**
Bartlett et al. [72]	Reduced-order electrochemical model for a composite electrode battery with solid particle and liquid sub-models **WHITE BOX**	SOH	• Chemistry: LMO-NMC (15 Ah) • The cells were cycled using the charge-depleting (CD) current profile defined by the US Advanced Battery Consortium	Loss of cyclable Li-ion that causes a shift of the normalized concentration operation ranges of the electrodes	≈85%	SOH estimation was performed on five different automotive cells tested at different conditions Mean estimate error: below 0.48 Ah
Li et al. [42]	Single particle-based degradation model **WHITE BOX**	SOH	• Chemistry: LFP (2.2 Ah) • Conditions shown in [73, 74]	• Cycle number • Temperature	≈76%	Error for predicted battery capacity fade RMSE: 10.3×10^{-3}
Gao et al. [38]	Order-reduced electrochemical model considering side reactions **WHITE BOX**	SOH	• Chemistry: NMC (26 Ah) • Ch: 1C (CCCV) protocol • Followed by a 30 min rest • Dch: 1C (CC) • Ambient temperature: 25°C	Capacity fade with the help of equilibrium electrode potential	60%	For cycles at 1C, 2C, and 3C Maximum error is mostly <2%
Lin et al. [58]	Fuzzy logic identification based on the closest normal distribution **BLACK BOX**	SOH	• Chemistry: LCO (3.7 V/ 2.37 Ah) • Ch: 0.5 C (CCCV protocol) • Dch: 0.2, 0.4, 0.6, 0.8, and 1C (CC) • Temp: 0–45°C	• Battery charging time • OCV difference between fully charged battery and with a load • Voltage difference between fully discharge and after resting for 1 min	≈70%	Average error of good diagnosis: 1.46%
Long et al. [74]	Autoregressive model and the improved particle swarm optimization algorithm **BLACK BOX**	RUL	• CALCE dataset: LCO (1.1 Ah) • Ch: 0.5C (CCCV protocol) • Dch: 0.5C (CC) • Ambient temperature	Capacity degradation	80% (defined threshold: 211 cycles)	RUL prediction difference at: Cycle 110: 26 cycles Cycle 140: 1 cycle Cycle 150: 0 cycle Cycle 190: 1 cycle
Zhang et al. [56]	Three-layer back propagation artificial neural network model **BLACK BOX**	SOH	• Batteries from Beijing Olympic EV bus	Internal resistance	Not reported. But they reach the 80% SOH from its use on second life	• Average absolute error 0.899 Ah • Capacity estimation error within 2.5%

Authors	Estimation method	Estimated variables	Experimental conditions*	Aging features employed for estimation	Minimum SOH reached	Reported error**
			• Internal resistance measure under (100 A, −200 A, and −300 A) pulsed current tests			
Zhou et al. [75]	Simple linear regression **BLACK BOX**	SOH	• Chemistry: LCO (1.1 Ah) • Ch: 0.5C (CCCV protocol) • Dch: 1C (CC)	Integral from voltage series between 3.85 and 4.3 time on CC charging phase	≈75%	Average R^2: 0.97 Average RMSE: 0.01
Cai et al. [60]	Support vector regression and genetic algorithm **BLACK BOX**	SOH	• LFP (3.3. V/2.5 Ah) • Load profile of primary frequency regulation • Ambient temperature: 25°C	Keen points in the voltage response under current pulse test	≈84%	RMSE for Cell 1: 19.12×10^{-3} Cell 2: 13.14×10^{-3}
Jiang et al. [51]	Incremental capacity analysis with multiple linear regression model and OLS estimation **BLACK BOX**	SOH	• LFP (60 Ah) obtained from a retired battery pack Load profiles of: • Frequency regulation application • Peak shaving application • Ambient temperature: 25°C	Evolution of normalized peaks of the incremental capacity curve	≈65%	Average errors for OLS regression: MAE (%):0.609 ME (%):1.226 RMSE: 0.589
Wu et al. [76]	Neural network model with a bat-based particle filter algorithm **BLACK BOX**	RUL	• CALCE dataset: LCO (1.1 Ah) • Ch: CCCV protocol • Dch: 1C (CC) • NASA dataset $LiCoO_2$ (2.1 Ah) Test (1) for periods of 5 min: • Ch: Series of random current • Dch: CC Test (2) 2A charging/discharging test after about 5 days	Cycle number or cycle time	80% (CALCE defined threshold: 602 cycles) (NAS2 defined threshold: 146.83 days)	• Capacity degradation fit: $R^2 > 0.98$ RMSE < 0.04 • RUL predictions: • For CALCE: AE: 2 cycles (at 500 cycles) • For NASA: AE: 2.19 days (at 100.02 days)
Quinard et al. [52]	Partial coulometric counter **BLACK BOX**	SOH	• LMO-LNO (3.75 V/65 Ah) • Full CC discharge at 1C forerun by a wake-up cycle (partial charge)	Partial capacity from a partial charge	≈45%	For partial counter: R^2: 0.69 Average AE: 1.6

Authors	Estimation method	Estimated variables	Experimental conditions*	Aging features employed for estimation	Minimum SOH reached	Reported error**
			• Sampling frequency: 10 Hz • Ambient temperature: 25°C			Maximum AE: 5.1 Estimated test time: 300 s
Casals et al. [77]	Aging model based on an equivalent electric circuit that simulates the battery's behavior **GRAY BOX**	SOH/RUL	Real demand area regulation profile from the Spanish operator "Red Eléctrica"given to a gas turbine power plant	Current (load profile) and temperature	Considering two SLB applications on providing area regulation service: Application 1: ≈51% Application 2: ≈46%	Considering two SLB applications on providing area regulation service: Application 1: deviation of 7.35% Application 2: deviation of 8.1%
Wei et al. [61]	Support vector regression-based state-space model, equivalent circuit, and particle filter **GRAY BOX**	RUL/SOH	• Gen 218,650-size LIBs • Ch CCCV: 1.5 A CC until 4.2 V and CV continue until 20 mA • Dch CCCV: 2 A CC until 2.7	Aging features extracted from CV protocol	≈65%	RMSE SOH SVR-PF [mΩ] #5 (5.1) #6 (8.7) #7(6.6) #18 (5.7) RUL prediction difference below 4 cycles
Tian et al. [69]	Online OCV estimation based on FOM and FFRELS **GRAY BOX**	SOH	Commercial NMC T: 10, 25, and 40°C • Ch: 1C • Dch: 1C, 2C, and 3C	ICA peaks	60%	Capacity fade error less than 3.1%
Hu et al. [71]	mMHE integrated to first-order RC ECM **GRAY BOX**	SOH	Panasonic NCR18650B (3.35 Ah) at 25°C with maximum voltage and current 5 V and 100 A, respectively	ECM parameters	Not reported	Relative error of capacity within 3%

**Conditions: Ch: charge conditions; Dch: discharge conditions; CCCV: constant current-constant voltage charging protocol.*
***Errors: AE: absolute error; MAE: mean absolute error; ME: maximum error; RMSE: root mean squared error.*

Table 1.
Comparative summary of SOH and RUL estimation methods.

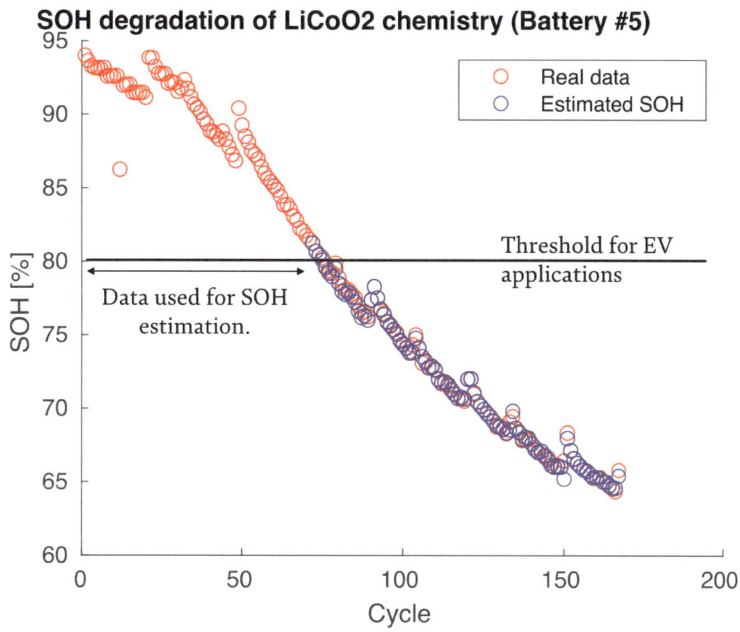

Figure 5.
SOH estimation results for battery #5 from NASA dataset [53]. Model constructed using data before battery reach the 80% of its nominal capacity (authors' results).

3.5 Brief discussion on the adaptability of EV estimation methods to SLBs estimation methods

As it has been discussed throughout this chapter, there is a lack of literature for SOH and RUL estimation methods validated for SLBs. In contrast, SOH and RUL variables have been extensively studied for first-life applications for EVs. Although some published works have developed approaches for diagnosis and prognostics of SLBs applied to real second-life scenarios, such as [26, 51, 52, 56, 77], we wanted to check if a SOH estimation method developed for EV application, designed for a SOH value of 80%, could be extended to SOH values below this threshold. Hence, the black-box method proposed by Zhou et al. [75] was used for this purpose. This method calculates the integral under the constant current section of a current–voltage curve, which was obtained using the constant current-constant voltage (CCCV) charging protocol as an aging feature. **Figure 5** shows the SOH estimation for a battery with $LiCoO_2$ chemistry until SOH values as low as 65%. An RMSE of 0.2140 was obtained. Therefore, the authors believe that SOH and RUL estimation methods commonly employed for electric vehicle applications could be extended to estimate these variables in SLBs. However, to guarantee a better accuracy, different battery degradation behaviors must be considered depending on the load profile for each future use.

4. Conclusions and final remarks

Electrochemical energy storage in the form of Li-ion batteries is proving to be a fundamental technology to catalyze an energy transition towards renewables and electric mobility. The EV worldwide fleet, and thus the amount of batteries, is expected to grow considerably in the following years. When EV batteries reach their end of life (SOH ≈ 80%), they can still store enough energy and can be used in other applications as second-life batteries. Otherwise, they would end up as waste. It is in this context, under a circular economy scenario, that retired EVs are regarded as a primary source of SLBs. In order to do this, an accurate estimation of the state of health and remaining useful life is crucial to determine if the battery is depleted and goes to recycling or if it may be used as a SLB. Thus, sophisticated SOH and RUL estimation methods are needed to guarantee the correct performance of SLBs in different applications.

In this review chapter, we classified these methods in three categories, namely, white-box, black-box, and gray-box, which refer to the level of theoretical or experimental knowledge needed to describe the aging process in batteries. Each category has its advantages and disadvantages, and its implementation will ultimately depend on the context it will be applied. White-box methods, which are usually employed in laboratory environments, are important because they elucidate what happens inside a battery in terms of aging/degradation, and, usually, the estimation errors are lower. However, they imply the use of complex physicochemical and mathematical models and require a higher computational cost. Black-box methods, commonly employed in commercial battery management systems, make use of mathematical and stochastic equations to establish correlations between intrinsic electrochemical mechanisms and external variables that can be easily measured. Although their computational cost is usually low, they need a high amount of data to establish these correlations. Finally, gray-box methods, which are hybrid prognostics between white- and black-box methods, are considered as a promising alternative for more accurate SOH/RUL estimation as they take into account both internal mechanisms of batteries and data-driven models.

In conclusion, although there is a lack of literature for SOH and RUL estimation methods for SLBs, extensive diagnostic and prognostic approaches have been developed for EV applications. The authors believe that some of these methods could be extended to estimate these variables in SLBs. However, to guarantee a better accuracy, different battery degradation behaviors must be considered depending on the energy loads of each future use. Nevertheless, batteries intended to be repurposed in second-life applications will have to compete, at the end of their first life, with improved battery technologies and chemistries that will be likely produced at lower costs in the near future.

Acknowledgements

This work was supported by Universidad Pontificia Bolivariana (UPB) in Colombia. Special acknowledgement goes to Seeding Labs and their Instrumental Access initiative which partially supported the foundation of the LIMAE research laboratory in UPB.

Author details

Simon Montoya-Bedoya[1], Laura A. Sabogal-Moncada[1*], Esteban Garcia-Tamayo[1,3,4] and Hader V. Martínez Tejada[2,3,4]

1 Research Laboratory in Materials for Energy (LIMAE), Department of Nanotechnology Engineering, UPB University, Medellín, Colombia

2 Research Laboratory in Materials for Energy (LIMAE), Department of Mechanical Engineering, UPB University, Medellín, Colombia

3 Grupo de Investigación sobre Nuevos Materiales (GINUMA), Department of Mechanical Engineering, UPB University, Medellín, Colombia

4 Grupo de Investigación Energía y Termodinámica (GET), Department of Mechanical Engineering, UPB University, Medellín, Colombia

*Address all correspondence to: laura.sabogal@upb.edu.co

References

[1] United Nations. The Sustainable Development Goals Report 2019. New York: United Nations Publ. issued by Dep. Econ. Soc. Aff.; 2019. p. 64

[2] Economic Commission for Latin America and the Caribbean and the Caribbean/Organisation for Economic Cooperation and Development. Emerging Challenges and Shifting Paradigms: New Perspectives on International Cooperation for Development. Santiago: United Nations; 2018

[3] IEA (International Energy Agency). CO_2 emissions by sector, World 1990–2017. In: CO_2 Emissions from Fuel Combustion 2019. 2019. ISBN: 978-92-64-32021-5

[4] REN21. Renewables 2019 Global Status Report. Renewable Energy Policy Network 21st Century; 2019. p. 336. ISBN: 978-3-9818911-7-1

[5] IEA (International Energy Agency). Electricity generation by source, World 1990–2017. In: Electricity Information 2019. 2019. ISBN: 978-92-64-98635-0

[6] United Nations. United Nations Framework on Climate Change. Adopt. Paris Agreement. FCCC/CP; 2015

[7] Deng D. Li-ion batteries: Basics, progress, and challenges. Energy Science & Engineering. 2015; **3**(5):385-418. DOI: 10.1002/ese3.95

[8] IRENA. Electricity Storage and Renewables: Costs and Markets to 2030. October. 2017. ISBN: 978-92-9260-038-9

[9] Garcia-Tamayo E. Advanced Thin Layer Deposition of Materials for Li-ion Batteries via Electrospray. TUDelft; 2014. DOI: 10.4233/uuid:bd76d1c8-448d-4f4a-b722-3f40c2810d3c

[10] World Economic Forum. A Vision for a Sustainable Battery Value Chain in 2030 Unlocking the Full Potential to Power Sustainable Development and Climate Change Mitigation. September 2019

[11] Vetter J et al. Ageing mechanisms in Li-ion batteries. Journal of Power Sources. Sep. 2005;**147**(1–2):269-281. DOI: 10.1016/j.jpowsour.2005.01.006

[12] Birkl CR, Roberts MR, McTurk E, Bruce PG, Howey DA. Degradation diagnostics for Li-ion cells. Journal of Power Sources. 2017;**341**:373-386. DOI: 10.1016/j.jpowsour.2016.12.011

[13] Aurbach D, Markovsky B, Weissman I, Levi E, Ein-Eli Y. On the correlation between surface chemistry and performance of graphite negative electrodes for Li ion batteries. Electrochimica Acta. Sep. 1999;**45**(1–2): 67-86. DOI: 10.1016/S0013-4686(99) 00194-2

[14] Agusdinata DB, Liu W, Eakin H, Romero H. Socio-environmental impacts of lithium mineral extraction: Towards a research agenda. Environmental Research Letters. 2018; **13**(12). DOI: 10.1088/1748-9326/aae9b1

[15] Acevedo M, Campagnol N, Hagenbruch T, Hoffman K, Lala A, Ramsbottom O. Lithium and Cobalt-a Tale of Two Commodities. Metals and Mining. McKinsey & Company; June 2018

[16] U.S. Geological Survey. Mineral Commodity Summaries 2019. 2019. DOI: 10.3133/70202434

[17] Harper G et al. Recycling Li-ion batteries from electric vehicles. Nature. Nov. 2019;**575**(7781):75-86. DOI: 10.1038/s41586-019-1682-5

[18] McKerracher C, et al. Electric Vehicle Outlook. 2019 [Online].

Available from: https://about.bnef.com/electric-vehicle-outlook/#toc-viewreport

[19] Miao Y, Hynan P, Von Jouanne A, Yokochi A. Current Li-ion battery technologies in electric vehicles and opportunities for advancements. Energies. 2019;**12**(6):1-20. DOI: 10.3390/en12061074ISBN:1254710221

[20] Harlow JE et al. A wide range of testing results on an excellent Li-ion cell chemistry to be used as benchmarks for new battery technologies. Journal of the Electrochemical Society. 2019; **166**(13):A3031-A3044. DOI: 10.1149/2.0981913jes

[21] Asbhby M, Polyblank J. White Paper Series. Cambridge, UK: Energy; 2012

[22] Melin HE. The Li-ion battery end-of-life market—A baseline study. Global Battery Alliance. 2018;**July**:1-11

[23] Xiong R, Shen W. Advanced Battery Management Technologies for Electric Vehicles. John Wiley & Sons, Inc.; 2019. ISBN: 978-1-119-48164-5

[24] Xing Y, Ma EWM, Tsui KL, Pecht M. Battery management systems in electric and hybrid vehicles. Energies. 2011;**4**(11):1840-1857. DOI: 10.3390/en4111840

[25] Yang F, Song X, Dong G, Tsui K-L. A coulombic efficiency-based model for prognostics and health estimation of Li-ion batteries. Energy. 2019;**171**: 1173-1182. DOI: 10.1016/j.energy.2019.01.083

[26] Casals LC, Amante García B, Canal C. Second life batteries lifespan: Rest of useful life and environmental analysis. Journal of Environmental Economics and Management. 2019;**232** (October):354-363. DOI: 10.1016/j.jenvman.2018.11.046

[27] Podias A et al. Sustainability assessment of second use applications of automotive batteries: Ageing of Li-ion battery cells in automotive and grid-scale applications. World Electric Vehicle Journal. 2018;**9**(2):24. DOI: 10.3390/wevj9020024

[28] Lipu MSH et al. A review of state of health and remaining useful life estimation methods for Li-ion battery in electric vehicles: Challenges and recommendations. Journal of Cleaner Production. 2018;**205**:115-133. DOI: 10.1016/j.jclepro.2018.09.065

[29] Lin C, Tang A, Wang W. A review of SOH estimation methods in Li-ion batteries for electric vehicle applications. Energy Procedia. 2015;**75**: 1920-1925. DOI: 10.1016/j.egypro.2015.07.199

[30] Barré A, Deguilhem B, Grolleau S, Gérard M, Suard F, Riu D. A review on Li-ion battery ageing mechanisms and estimations for automotive applications. Journal of Power Sources. 2013;**241**: 680-689. DOI: 10.1016/j.jpowsour.2013.05.040

[31] Ungurean L, Cârstoiu G, Micea MV, Groza V. Battery state of health estimation: A structured review of models, methods and commercial devices. International Journal of Energy Research. 2017;**41**(2):151-181. DOI: 10.1002/er.3598

[32] Berecibar M, Gandiaga I, Villarreal I, Omar N, Van Mierlo J, Van den Bossche P. Critical review of state of health estimation methods of Li-ion batteries for real applications. Renewable and Sustainable Energy Reviews. 2016;**56**:572-587. DOI: 10.1016/j.rser.2015.11.042

[33] Saidani F, Hutter FX, Scurtu R-G, Braunwarth W, Burghartz JN. Li-ion battery models: A comparative study and a model-based powerline communication. Advances in Radio

Science. Sep. 2017;**15**:83-91. DOI: 10.5194/ars-15-83-2017

[34] Liao L, Kottig F. Review of hybrid prognostics approaches for remaining useful life prediction of engineered systems, and an application to battery life prediction. IEEE Transactions on Reliability. Mar. 2014;**63**(1):191-207. DOI: 10.1109/TR.2014.2299152

[35] Xia L, Najafi E, Li Z, Bergveld HJ, Donkers MCF. A computationally efficient implementation of a full and reduced-order electrochemistry-based model for Li-ion batteries. Applied Energy. 2017;**208**(May):1285-1296. DOI: 10.1016/j.apenergy.2017.09.025

[36] Fu R, Choe SY, Agubra V, Fergus J. Modeling of degradation effects considering side reactions for a pouch type Li-ion polymer battery with carbon anode. Journal of Power Sources. 2014; **261**:120-135. DOI: 10.1016/j.jpowsour.2014.03.045

[37] Canals Casals L. Modelling Li-ion battery aging for second life business models TDX (Tesis Dr. en Xarxa). February 2016

[38] Gao Y, Zhang X, Yang J, Guo B. Estimation of state-of-charge and state-of-health for Li-ion degraded battery considering side reactions. Journal of the Electrochemical Society. 2018;**165**(16): A4018-A4026. DOI: 10.1149/2.0981816jes

[39] Lotfi N, Li J, Landers RG, Park J. Li-ion battery state of health estimation based on an improved single particle model. Proceedings of the American Control Conference. 2017:86-91. DOI: 10.23919/ACC.2017.7962935ISBN: 9781509059928

[40] Rechkemmer SK, Zang X, Zhang W, Sawodny O. Empirical Li-ion aging model derived from single particle model. Journal of Energy Storage. 2019; **21**(January):773-786. DOI: 10.1016/j.est.2019.01.005

[41] Li J, Lotfi N, Landers RG, Park J. A single particle model for Li-ion batteries with electrolyte and stress-enhanced diffusion physics. Journal of the Electrochemical Society. 2017;**164**(4): A874-A883. DOI: 10.1149/2.1541704jes

[42] Li J, Adewuyi K, Lotfi N, Landers RG, Park J. A single particle model with chemical/mechanical degradation physics for Li-ion battery state of health (SOH) estimation. Applied Energy. Feb. 2018;**212**: 1178-1190. DOI: 10.1016/j.apenergy.2018.01.011

[43] Groenewald J, Grandjean T, Marco J. Accelerated energy capacity measurement of Li-ion cells to support future circular economy strategies for electric vehicles. Renewable and Sustainable Energy Reviews. 2017;**69**: 98-111. DOI: 10.1016/j.rser.2016.11.017

[44] Grandjean T, Groenewald J, McGordon A, Widanage W, Marco J. Accelerated internal resistance measurements of Li-ion cells to support future end-of-life strategies for electric vehicles. Batteries. Oct. 2018;**4**(4):49. DOI: 10.3390/batteries4040049

[45] Hossain E, Murtaugh D, Mody J, Faruque HMR, Sunny MSH, Mohammad N. A comprehensive review on second-life batteries: Current state, manufacturing considerations, applications, impacts, barriers potential solutions, business strategies, and policies. IEEE Access. 2019;7: 73215-73252. DOI: 10.1109/ACCESS.2019.2917859

[46] Song Z, Feng S, Zhang L, Hu Z, Hu X, Yao R. Economy analysis of second-life battery in wind power systems considering battery degradation in dynamic processes: Real case scenarios. Applied Energy. 2019;**251**. DOI: 10.1016/j.apenergy.2019.113411

[47] Schuster SF et al. Nonlinear aging characteristics of Li-ion cells under different operational conditions. Journal of Energy Storage. 2015;**1**(1):44-53. DOI: 10.1016/j.est.2015.05.003

[48] Baumhöfer T, Brühl M, Rothgang S, Sauer DU. Production caused variation in capacity aging trend and correlation to initial cell performance. Journal of Power Sources. 2014;**247**:332-338. DOI: 10.1016/j.jpowsour.2013.08.108

[49] Wu L, Fu X, Guan Y. Review of the remaining useful life prognostics of vehicle Li-ion batteries using data-driven methodologies. Applied Science. 2016;**6**(6). DOI: 10.3390/app6060166

[50] Pan H, Lü Z, Wang H, Wei H, Chen L. Novel battery state-of-health online estimation method using multiple health indicators and an extreme learning machine. Energy. 2018;**160**: 466-477. DOI: 10.1016/j.energy.2018.06.220

[51] Jiang Y, Jiang J, Zhang C, Zhang W, Gao Y, Li N. State of health estimation of second-life $LiFePO_4$ batteries for energy storage applications. Journal of Cleaner Production. 2018;**205**:754-762. DOI: 10.1016/j.jclepro.2018.09.149

[52] Quinard H, Redondo-Iglesias E, Pelissier S, Venet P. Fast electrical characterizations of high-energy second life Li-ion batteries for embedded and stationary applications. Batteries. 2019;**5**(1):33. DOI: 10.3390/batteries5010033

[53] B. Saha and K. Goebel, Battery Data Set. NASA Ames Prognostics Data Repository, Moffett Field, CA, 2007. Available from: http://ti.arc.nasa.gov/project/prognostic-data-repository [Accessed: 05 August 2019]

[54] Birkl CR. Diagnosis and Prognosis of Degradation in Li-ion Batteries. University of Oxford; 2017. DOI: 10.5287/bodleian:KO2kdmYGg

[55] Zhou Y, Huang M. On-board capacity estimation of Li-ion batteries based on charge phase. Journal of Electrical Engineering and Technology. 2018;**13**(2):733-741. DOI: 10.5370/JEET.2018.13.2.733

[56] Zhang C, Jiang J, Zhang W, Wang Y, Sharkh S, Xiong R. A novel data-driven fast capacity estimation of spent electric vehicle Li-ion batteries. Energies. 2014;**7**(12):8076-8094. DOI: 10.3390/en7128076

[57] Lee S, Cui H, Rezvanizaniani M, Ni J. Battery prognostics: SoC and SoH prediction. In: ASME 2012 International Manufacturing Science and Engineering Conference. 2012. pp. 689-695. DOI: 10.1115/MSEC2012-7345ISBN:978-0-7918-5499-0

[58] Lin H-T, Liang T-J, Chen S-M. The state-of-health diagnosis of Li-Co batteries with fuzzy identification. In: Proceedings of The 7th International Power Electronics and Motion Control Conference. 2012. pp. 2678-2682. DOI: 10.1109/IPEMC.2012.6259285ISBN: 978-1-4577-2085-7

[59] Severson KA et al. Data-driven prediction of battery cycle life before capacity degradation. Nature Energy. 2019;**4**(5):383-391. DOI: 10.1038/s41560-019-0356-8

[60] Cai L, Meng J, Stroe D-I, Luo G, Teodorescu R. An evolutionary framework for Li-ion battery state of health estimation. Journal of Power Sources. 2018;**412**:615-622. DOI: 10.1016/j.jpowsour.2018.12.001

[61] Wei J, Dong G, Chen Z. Remaining useful life prediction and state of health diagnosis for Li-ion batteries using particle filter and support vector regression. IEEE Transactions on Industrial Electronics. 2018;**65**(7): 5634-5643. DOI: 10.1109/TIE.2017.2782224

[62] Tang X, Zou C, Yao K, Lu J, Xia Y, Gao F. Aging trajectory prediction for Li-ion batteries via model migration and Bayesian Monte Carlo method. Applied Energy. 2019;**254**:113591. DOI: 10.1016/j.apenergy.2019.113591

[63] Ma C et al. State of health prediction for Li-ion batteries using multiple-view feature fusion and support vector regression ensemble. International Journal of Machine Learning and Cybernetics. 2019;**10**(9):2269-2282. DOI: 10.1007/s13042-018-0865-y

[64] Ma G, Zhang Y, Cheng C, Zhou B, Hu P, Yuan Y. Remaining useful life prediction of Li-ion batteries based on false nearest neighbors and a hybrid neural network. Applied Energy. 2019; **253**:113626. DOI: 10.1016/j.apenergy.2019.113626

[65] Liao L, Köttig F. Review of hybrid prognostics approaches for remaining useful life prediction of engineered systems, and an application to battery life prediction. IEEE Transactions on Reliability. 2014;**63**(1):191-207. DOI: 10.1109/TR.2014.2299152

[66] Li X, Wang Q, Yang Y, Kang J. Correlation between capacity loss and measurable parameters of Li-ion batteries. International Journal of Electrical Power & Energy Systems. 2019;**110**(January):819-826. DOI: 10.1016/j.ijepes.2019.03.046

[67] Bi J, Zhang T, Yu H, Kang Y. State-of-health estimation of Li-ion battery packs in electric vehicles based on genetic resampling particle filter. Applied Energy. 2016;**182**:558-568. DOI: 10.1016/j.apenergy.2016.08.138

[68] Shi E, Xia F, Peng D, Li L, Wang X, Yu B. State-of-health estimation for lithium battery in electric vehicles based on improved unscented particle filter. Journal of Renewable and Sustainable Energy. 2019;**11**(2):024101. DOI: 10.1063/1.5065477

[69] Tian J, Xiong R, Yu Q. Fractional-order model-based incremental capacity analysis for degradation state recognition of Li-ion batteries. IEEE Transactions on Industrial Electronics. 2019;**66**(2):1576-1584. DOI: 10.1109/TIE.2018.2798606

[70] Guo Q et al. Estimation of electric vehicle battery state of health based on relative state of health evaluation. In: Proceedings of the 2017 IEEE 2nd Advanced Information Technology, Electronic and Automation Control Conference, IAEAC 2017, vol. 1. 2017. pp. 1998-2002. DOI: 10.1109/IAEAC.2017.8054365ISBN: 9781467389778

[71] Hu X, Jiang H, Feng F, Liu B. An enhanced multi-state estimation hierarchy for advanced Li-ion battery management. Applied Energy. 2020; **257**:114019. DOI: 10.1016/j.apenergy.2019.114019

[72] Bartlett A, Marcicki J, Onori S, Rizzoni G, Yang XG, Miller T. Electrochemical model-based state of charge and capacity estimation for a composite electrode Li-ion battery. IEEE Transactions on Control Systems Technology. 2015:1-1. DOI: 10.1109/TCST.2015.2446947

[73] Liu P et al. Aging mechanisms of $LiFePO_4$ batteries deduced by electrochemical and structural analyses. Journal of Electrochemical Society. 2010;**157**(4):A499. DOI: 10.1149/1.3294790

[74] Long B, Xian W, Jiang L, Liu Z. An improved autoregressive model by particle swarm optimization for prognostics of Li-ion batteries. Microelectronics and Reliability. 2013; **53**(6):821-831. DOI: 10.1016/j.microrel.2013.01.006

[75] Zhou Y, Huang M, Pecht M. An online state of health estimation method for Li-ion batteries based on integrated

voltage. In: 2018 IEEE International Conference on Prognostics and Health Management, ICPHM 2018, no. 2015. 2018. pp. 1-5. DOI: 10.1109/ICPHM.2018.8448947ISBN: 9781538611647

[76] Wu Y, Li W, Wang Y, Zhang K. Remaining useful life prediction of Li-ion batteries using neural network and bat-based particle filter. IEEE Access. 2019;7:54843-54854. DOI: 10.1109/ACCESS.2019.2913163

[77] Casals LC, García BA. Second-life batteries on a gas turbine power plant to provide area regulation services. Batteries. 2017;**3**(1). DOI: 10.3390/batteries3010010

Chapter 5

Leveraging Integrated Model-Based Approaches to Unlock Bioenergy Potentials in Enhancing Green Energy and Environment

Fabrice Abunde Neba, Prince Agyemang, Yahaya D. Ndam, Endene Emmanuel, Eyong G. Ndip and Razak Seidu

Abstract

In the quest for a green economy, bioenergy has become a central component due to its ability to minimize depletion of natural energy resources and enhance environmental sustainability. However, the integration of bioenergy for a green economy has often led to policy resistance, the tendency for solutions to cause disastrous side effects on other aspects of the system that were not envisaged. The use of integrated model-based approaches for selection, design, and analysis of technological alternatives for bioenergy production would significantly enhance the systems' sustainability by optimizing design and operation, improving growth and profitability, and enabling a more synergistic interaction between the engineering and the macroeconomic aspects of bioenergy production systems. This chapter is designed to develop model-based methodological frameworks that will support sustainable decision making by all stakeholders involved in the design, operation, and commercialization of bioenergy production systems. Practical case studies are presented for bioethanol, biomethane, and synthetic gas production.

Keywords: system thinking, model identification and analysis, bioreactor synthesis, performance targeting, and economics

1. Introduction

1.1 Bioenergy as a source of sustainable energy

Increasing concerns about depletion of natural resources, precarious nature of waste management and sanitation challenges, as well as environmental deterioration and climate change, have led to a growing interest by many countries to switch to renewable energy technologies. Consequently, the last two decades have seen a rapid implementation of new renewable energy systems, followed by integration of renewable energy into plants where fossil fuels exist. Amongst the existing renewable energy technologies, bioenergy systems are of special significance, because in addition to being able to generate renewable energy, these systems also breakdown pollutants as well as recycle valuable nutrients found in organic waste [1]. Recent studies have confirmed that the bioenergy technology is robust and offers a great potential not only to reduce energy poverty through the provision of green energy but also enhance a green environment by reducing emissions associated with poor waste management [2–4]. According to one estimate on a bioenergy system, the anaerobic digestion technology, co-digestion of wastewater in a decentralized treatment plant with food wastes could allow the generation of 0.9 kWh electricity per person per day, leaving the nutrients as part of organic matter intact for agricultural use [5]. The recognition of the advantages of bioenergy systems in complying with the progressively more restrictive environmental requirements has led to an increased development and use of new bioenergy technologies, some of which include: sugar fermentation for bioethanol production, anaerobic digestion for biogas generation, pyrolysis for bio-oil production, microbial fuels cells for electricity generation, transesterification for biodiesel production, gasification for syngas production, etc. [6, 7].

Special challenges arise when attempting to implement a bioenergy technology for renewable energy generation in a given community. Firstly, assessing and selecting the optimal technological alternatives that meet social, economic, and environmental sustainability standards is a challenging task. This is because the successful operation of bioenergy systems depends on the availability of a sustainable supply of feedstock, requiring tradeoffs to be made, on whether to use feedstocks and other utilities for bioenergy generation or to channel theses inputs to other industrial sectors requiring the same feedstocks and utility. In addition, bioenergy systems have specific characteristics, making them more adapted to specific feedstocks than others. Secondly, after knowing the technology to implement, challenges often arise from deciding on an optimal spatio-temporal strategy to implement the technology. A long-term perspective is needed to account for the spatio-temporal impact of the bioenergy system on the community to ensure that the system does on result in disastrous side-effects. Some systems might be reliable over the short term or in a given location but pose significant negative effects in the long-term or other locations. It is highly important to use systematic model based techniques to understand the possible impacts of a given bioenergy system over time horizons that span from months to years, and determine the optimal implementation strategies, which maximize the positive effects and minimize the unwanted side-effects. Thirdly, wouldn't it be surprising if the authors state that getting the right technology and the right implementation strategy doesn't guarantee successful operation? Optimal operation of bioenergy systems requires optimal process configurations that ensure process stability, as well as maximize yield and productivity to ensure economic sustainability of the plant. Systematic mathematical and computational techniques are required for process modelling and simulation aimed at synthesizing optimal plant configurations well adapted to the specific feedstock characteristics of interest. Finally, after obtaining optimal process configurations, it is important to now shift the focus away from the technological solution and placing the focus on the practical considerations required for construction and installation of the technology vis-à-vis the required performance or need. This is highly important because the right technology, with optimal implementation strategy and optimal process configurations, can fail because of wrong equipment characteristics. For the same bioenergy technology, the choice of equipment components required for a rural community in a developing country would not be the same as that required in an urban setting. In addition, choice of equipment characteristics plays a significant role in the cost of installation, and there have been cases where projects have failed to get to completion due to high cost required for implementation. Systematic model-based techniques are again required to understand how every specific component of the bioenergy system provides the value with regards to meeting the overall needs of the community. The objective here is to improve the economic value of the product by examining each component to determine how many functions that component performs, and the cost contributions of those functions. Systems components with high cost-function ratios are identified as opportunities for further investigation and improvement. The authors will like to mention here that a green economy cannot be achieved without changing the way we design and implement our solutions. However, international discussions on sustainable development have always focused on new technologies that can guarantee sustainability and ignored strategies or tools that can be used to design and implement technological solutions optimally. This chapter is therefore designed to fill this gap and provide a series of model-inspired tools, which can be used to enhance the successful implementation of bioenergy technologies.

1.2 Model-based techniques for sustainable bioenergy systems

When confronted by any complex system, with things about it that you're dissatisfied with (environmental pollution and climate change) and anxious to fix (such as using bioenergy to enhance a green economy), you cannot just step in and set about fixing with much hope of helping. This realization is one of the sore discouragements of our century. You cannot meddle with one part of a complex system from the outside without the almost certain risk of setting off disastrous events that you hadn't counted on in other, remote parts. If you want to fix something, you are first obliged to understand the underlying dynamics of the whole system. Intervening is a way of causing trouble. This is because positively intended solutions to real problems have often led to policy resistance, the tendency for interventions to be delayed, diluted, or defeated by the response of the system to the intervention itself. Considering the complex nature of the economy, model-based techniques, which simultaneously considers all degrees of freedom (the efficiency of the solution as well as the quantitative social economic and environmental impact both in time and space) by employing mathematical models becomes undoubtedly the strategy of choice. To understand the problems better and make good decisions, appropriate analysis tools are required. Previous analysis has tended to use 'soft' approaches, which do not require a knowledge of mathematical or computational techniques. These approaches can often be complementary to the techniques presented here. However, the use of mathematical and computational methods can be advantageous, due to the complexity and interactive nature of many of the problems involved and can, for instance, support decision making and making trade-offs in complex problems. However, many researchers are unfamiliar with the range of analytical, mathematical, and computational methods that could be applied in this area. Therefore, they are not able to take advantage of the full range of available methods in their research or analysis. This book chapter aims to fill this gap by providing both a basic introduction and advanced technical details of some of the available mathematical and computing methods, as well as illustrating their use through case studies and examples.

The methods presented here are aimed specifically at sustainable deployment of bioenergy technologies into production, and the case studies and examples are all in this area, but they have a wide range of other application areas, including in economics, medicine, and control systems. The techniques presented include:

- Systems theory and methodologies for structuring complex sustainable development problems to make it easier to obtain a solution to them.

- Optimization and decision-making techniques to support policy formulation and other decision applications.
- Attainable region technique for performance targeting, synthesis as well as analysis of process configurations required to operate bioenergy systems
- Functional analysis system technique, use to ensure that the engineering systems are designed and constructed with minimal cost, and strongly align the needs of the community using the system.

2. Theoretical background

2.1 Conceptual framework of an integrated model-based approach

The conceptual model-based framework that can be used for the implementation of a given bioenergy technology in a community is illustrated in **Figure 1**, consisting of five major stages. Stage one involves deciding on the type of bioenergy to install. This decision making in energy supply is influenced by factors such as social, economic, environmental, political, and technical impact, making it helpful in developing a sustainable solution to the local community. Due to the difficulty in complex interactions between the aforementioned factors, the Multicriteria Decision Method (MCDM) is employed. This provides an approach that eliminates the challenges by developing evaluation criteria and methods that reliably measure sustainability, leading to the selection of an appropriate bioenergy system for the community. The next stage involves the use of system dynamic modelling to devise an implementation strategy for the proposed bioenergy technology from stage one. This stage involves the development of linear and non-linear mathematical models for the underlying mechanism of the system and evaluating the dynamic behaviors to identify policy resistance and any human decisions that can exacerbate perturbations. The strength of the technique is that it helps to minimize unforeseen side effects and generate a forecast to determine future side effects, aiding in the

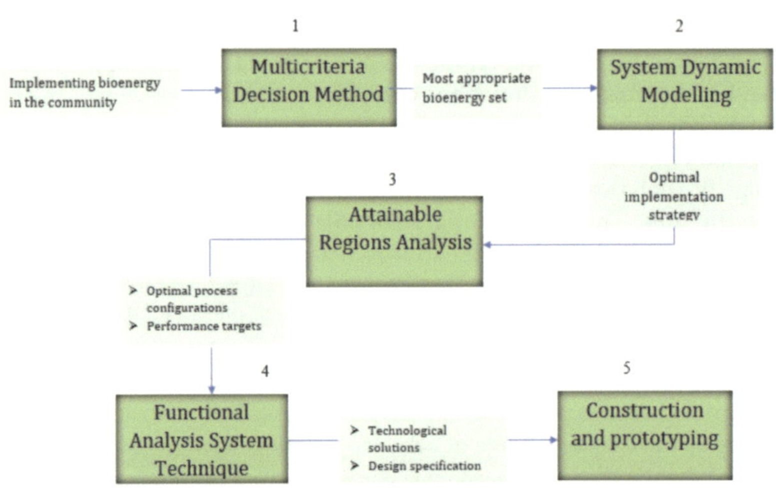

Figure 1.
Model-based framework for bioenergy systems implementation in rural communities.

selection of an optimal implementation strategy for the system. In stage three, the concept of attainable regions is used for the design of optimal process configurations. The AR technique is a systematic approach to process synthesis, which integrates elements of geometry and mathematical optimization to understand how systems can be designed and improve. The power of the AR technique is that all possible states, for all possible bioreactor configurations, are first determined [8–10]. The AR can be constructed after specifying the geometric space, kinetic models, and the feed conditions; then, the appropriate objective functions can be overlaid on the AR boundary to identify the optimal operating points and associated process configurations [8]. Once the optimal process configuration has been identified, the next stage is to analyze the design process adequately. This is achieved by employing the functional analysis method, a technique that provides technological solutions and design specifications which permit the satisfaction of the principal and constraint function. Interesting, the technique provides an effective method in improving the quality and performance as well as minimizing the cost of the proposed solution [11]. Finally, the proposed solution can be constructed. It is worth noting that it is not the intention of the authors to discuss the construction phase, although it is mentioned.

2.2 Modeling concepts

2.2.1 Multicriteria decision making

The implementation of successful sustainable energy solutions in local communities consist of a balance of social, economic, technical, and environmental aspects [12]. In that, the energy generated should: (1) be within an environmental tolerance limit, (2) generate employment opportunities for the locals, thereby improving their income and contribute to the regional or national economy, (3) meet the energy demand using available feedstock in the local community [13, 14]. These factors cannot be achieved without a systematic methodological framework that simultaneously considers all the degrees of freedom associated with the bioenergy system. This is because complex interactions exist between these factors, making the decision on the type of bioenergy system difficult. Therefore, sustainable decision making should integrate MCDM tool for successful bioenergy installation as this would guarantee the potentials for an increased standard of living as well as social and economic stability. The technique requires deciding on the type of decision model to employ and developing an evaluation criterion that reliably selects the best bioenergy alternative based on the aforementioned factors. **Table 1** illustrates the different categories of classifying MCDM and their respective methodologies.

2.2.2 System dynamics modeling

A system is a set of interrelated elements, where any change in any element affects the set as a whole. Only elements directly or indirectly related to the problem form the system under study. To study a system, we must know the elements that make it up and the relationships between them. A strong focus must be geared towards understanding the characteristics of its constituents and the nature of the relationship that exists between them. This is necessitated because, more often than not, positively intended solutions to real-life problems have often led to unwanted side effects that were not envisaged [16]. Therefore considering the complex nature of systems, a system thinking approach that simultaneously considers all degrees of freedom of the problem is undoubtedly the strategy of choice. System thinking, also known as system dynamics modelling is a scientific framework for addressing

Categories	Methodology
Multi-attribute utility and value theory	AHP/ANP
Multi-objective mathematical programming	• Constrain programming • Linear programming • Goal programming
Non-classical method	Fuzzy set methodology
Elementary aggregation method	• Weighted sum method • Weighted product method
Complex aggregation method	ASPID
Distance-to-target approach	• TOPSIS • Grey Relational Analysis • Data Enveloping Analysis
Direct ranking (high dependence on decision-maker)	• Stepwise expert judgment • Delphi • Scoring method
Outranking method	• ELECTRE I, IS, II, III • PROMETHEE I, II

Compiled from Refs. [13–15].

Table 1.
Categories for classifying MCDM methodologies.

complex, nonlinear feedback systems [17]. The strength of this technique is that it provides an opportunity to understand the dynamic behavior of the system under study and generate useful information that affects policy evaluations.

2.2.3 Attainable regions (performance targeting and equipment design)

Attainable region (AR) is an approach to graphical/geometric optimization of bioreactor network synthesis. The technique originated from the work of Horn, who defined the AR as the set of all possible values of the outlet stream variables, which can be reached by any possible (physically realizable) steady-state reactor system from a given feed stream using only the processes of reaction and mixing. The technique has been used in the synthesis of isothermal reactor networks [18], synthesis and design of biogas digester structures [19–22], classical variations and dynamic problem synthesis, optimal batch distillation for reduced energy consumption [23, 24], and in analysis to optimize particle breakage in ball mills [25]. Interestingly, in recent publications by Abunde et al., the concept of AR has been extended to include Self-optimizing attainable regions for the design of anaerobic digesters [21]. This is the first of its kind and saught to addresses the design of anaerobic digesters in situations where reliable kinetic coefficients are unavailable. The technique offers exciting possibilities for process synthesis seeing the countless opportunities it holds to address reactor network synthesis problems. More importantly, there are speculations of an extension of the concept to the design of dryers and distillation columns. Other future studies could look at how self-optimizing AR for design could be integrated with self-optimizing controllers to achieve optimality in processes. The strength of the AR approach is that it simultaneously considers all possible outputs for all possible process configurations, by interpreting the process as a geometric object that defines the limits of achievability without having to enumerate all reactor configurations explicitly [8].

2.2.4 Functional analysis (need analysis and technological features, design specification)

Function analysis, a vital component of value analysis, is a technique employed during the design and construction stage to assess a product's function to eliminate components that neither contribute to the quality nor improve the efficiency product. This technique provides an assessment of the proposed technology from different perspectives to adequately identify possible rooms of improvement [11]. The advantage of the technique is that it allows a transition from a focus on the expected solution to a problem to the appropriate and desired performance needs of the product.

3. Workflow of concepts and integrated model-based methodology

3.1 Workflow of methodology

This section discusses in detail, a step by step approach to how the techniques presented in the conceptual framework are deployed.

3.2 Overview of MCDM methodology

There are several MCDM tools to deploy in selecting an appropriate bioenergy technology. The decision on the tool to employ at this stage is vital and requires selecting the best alternative, which is a difficult decision since there exists characteristics that are peculiar to each technology.

It is expedient for readers to note that not all MCDM methods presented in Section 2.2.1 are same. This is because some methods incorporate more features than others that are rather limited from different perspective [26]. Moreover, the choice of method is usually dependent on the decision makers' knowledge of the techniques and the availability of software that support the method ([14, 27]). In this regard, most multicriteria decision problems are adjusted to suite a particular method. However, subjective and objective MCDM tools such as Analytical Hierarchy Process (AHP) and Technique for order preference by similarity to ideal solution (TOPSIS), respectively, are often used in sustainable energy decisions. AHP provides a very simple and flexible model for a problem and is useful in achieving a consensus in cases where there are multiple conflicting criteria. However, its inability to capture uncertainties and determine alternative ratings in decision making is complimented by TOPSIS, making the use of an integrated AHP-TOPSIS technique a more robust approach to decision making.

It is interesting to note that there exist frameworks that aid the selection of a decision-making tool as presented by Watróbski et al [27], that links a decision-making situation to the most suitable multicriteria decision method. However, this presents an inexhaustive, detailed and nearly impossible approach that takes into consideration all decision dimension, not to mention the extensive number and variety of methods available (in the reported presented by Watróbski et al, over 56 decision making tools exist). For this reason the authors focused on AHP and TOPSIS which has been successfully deployed by Akash et al., for the successful selection of an electric power plant [28] and Mohsen et al., in evaluation of an electric heating system [29]. It is also worth noting that it is not the intention of the authors to describe how to select a tool but rather to demonstrate how model-based techniques can be used to select an optimal bioenergy for a community. Interested readers can resort to the referenced material in this section.

AHP is a multi-level structured technique that presents a comprehensive framework for determining the different alternative solutions for a certain problem [30]. The technique was first introduced by Saaty in 1980 and is described in the following:

- The first step involves developing a hierarchy structure that describes the goal, alternatives, criteria, and sub-criteria for evaluation.

- Pair-wise comparison for the criteria and alternatives with respect to the goal (objective) is established to extract the decision matrices using a nine-point scale. Comparing an attribute to itself is assigned a value of 1 so that if an n given criteria matrix is constructed at any given level, the diagonal entries will all be 1. The value 1 also signifies, equal relevance of attributes. The numbers 3, 5, 7, and 9 correspond to "moderate importance," "strong importance," "very strong importance," and "absolute importance", respectively. It is important to note that the length of the pair-wise matrix is equivalent to the number of attributes.

- The pair-wise comparison procedure is repeated for each criterion, and then the priority of alternatives is acquired by accumulating the weights. Statistical techniques such as arithmetic mean method, characteristic root method, and least square method can be employed to accumulate the weights. Adopting an arithmetic sum approach, a vector $W = [W_1, W_2, \ldots W_N]$ is constructed to represents the weight of each criterion in a pair-wise comparison matrix A. each element in column j of matrix A is divided by the sum of entries in the j column. This step generates a new matrix called the normalized matrix (A_{norm}).

- The final step involves making a decision based on the priorities set, but before that, the normalized matrix is subjected to a consistency check to evaluate whether the comparison made was sound. The check involves determining the maximum Eigen values and consistency index using Eqs. (1) and (2), respectively. One advantage of the consistency ratio is that it eliminates the problem of disagreements in individual judgments.

$$\lambda_{max} = 1/n \sum_{i=1}^{n} \frac{i^{th} entry\ in\ AW^T}{i^{th}\ entry\ in\ W^T} \tag{1}$$

where λ_{max}, maximum Eigen value; n, number of attributes; A, pairwise comparison matrix; W, the estimate of the decision-maker's weight.

Nevertheless, the consistency is checked by comparing the consistency Index (CI) to the Random Index (RI) for the appropriate value of n, used in decision-making [30]. If (CI/RI) $<$ 0.10, the degree of consistency is satisfactory, but if (CI/RI) $>$ 0.10, serious inconsistencies may exist, and the results produced by AHP may not be meaningful.

$$CI = \frac{\lambda_{max} - n}{n - 1} \tag{2}$$

where the variables have their usual meaning.

TOPSIS selects the best alternative based on their geometric distance from the positive or negative ideal solution. According to the technique, the best alternative from the positive ideal solution has the shortest geometric distance, while the negative ideal solution has the longest geometric distance. Assuming for the bioenergy system understudy, we have m alternatives, n number of attributes, and the score of each alternative with respect to each criterion is known, the following steps could be followed in order to rank each alternative.

Step 1: Construct the normalized decision matrix

In this step, the different attributes dimensions are transformed into non-dimensional attribute, to allow comparison across the attributes. Using the method represented in Eq. (3), the matrix $(x_{ij})_{m\times n}$ is normalized to $R = (r_{ij})_{m\times n}$ which takes the form shown below:

$$r_{ij} = \frac{x_{ij}}{\sqrt{\sum_{j=1}^{m} x_{ij}^2}} \qquad i = 1, 2, \ldots, m; j = 1, 2, \ldots, n$$
$$R = \begin{pmatrix} r_{11} & \cdots & r_{1n} \\ \vdots & \ddots & \vdots \\ r_{m1} & \cdots & r_{mn} \end{pmatrix} \tag{3}$$

Step 2: Construct the weighted normalized decision matrix

With the normalized decision matrix (R) computed from the previous step, the weighted matrix W from the AHP method is integrated into the R. This results in a matrix that is computed by multiplying each column of R with its associated weighted matrix W as represented in Eq. (4).

$$V_{ij} = w_j \times r_{ij} \; where \; i = 1, 2, \ldots, n \tag{4}$$

This computation results in a new matrix V, which is represented below

$$V = \begin{bmatrix} v_{11} & \cdots & v_{1n} \\ \vdots & \ddots & \vdots \\ v_{m1} & \cdots & v_{mn} \end{bmatrix} = \begin{bmatrix} w_1 r_{11} & \cdots & w_n r_{1n} \\ \vdots & \ddots & \vdots \\ w_1 r_{m1} & \cdots & w_n r_{mn} \end{bmatrix}$$

Step 3: Determine the ideal and negative ideal solutions

In this process, two artificial alternatives A^* (the ideal alternative) and A^- (the negative ideal alternative) are defined as:

$$A^* = \{v_1^*, v_2^*, \ldots, v_n^*\} = \{(max\ _j v_{ij} | i \in I'), (min\ _j v_{ij} | i \in I'')\}, i = 1, 2, \ldots, m; j = 1, 2, \ldots, n.$$

$$A^- = \{v_1^-, v_2^-, \ldots, v_n^-\} = \{(min\ _j v_{ij} | i \in I'), (max\ _j v_{ij} | i \in I'')\}, i = 1, 2, \ldots, m; j = 1, 2, \ldots, n.$$

where I' is related to benefit attributes and I'' is related to cost attributes

Step 4: Achieve the remoteness of all choices from A^+ and A^-

In the process, the separation measurement is done by calculating the distance between each alternative in V and the ideal vector A^* using the Euclidean distant which is given as Eqs. (5) and (6)

$$D_i^+ = \sqrt{\sum_{j=1}^{n} \left(v_{ij} - v_j^+\right)^2} \; i = 1, 2, \ldots, m \tag{5}$$

$$D_i^- = \sqrt{\sum_{j=1}^{n} \left(v_{ij} - v_j^-\right)^2} \; i = 1, 2, \ldots, m \tag{6}$$

where D_i^+ and D_i^- are the Euclidean distance from the ideal best and ideal worst, respectively.

At the end of this, two quantities namely D_i and S_j for each alternative has been counted, representing the distance between each alternative and both (the ideal and the negative ideal).

Step 5: Determine the relative closeness to the ideal solution using Eq. (7).

$$CC_i^* = \frac{D_i^-}{D_i^- + D_i^+} \quad i = 1, 2, \ldots, m \tag{7}$$

where CC_i^* is the performance score.

Step 6: Rank the alternatives according to relative closeness to ideal solution

The set of the alternative Ai can now be ranked according to the descending order of CC_i^*, the highest value, the better performance. **Figure 2** represents an integrated AHP-TOPSIS for multicriteria decision making.

From **Figure 2**, the AHP is used to determine the weight of each criterion, while the TOPSIS is applied to achieve the final ranking of the alternative bioenergy technology closest to the ideal solution.

3.3 Overview of system dynamics modeling methodology

System dynamics deals with feedback and delays that affect system behavior over time. The power of the technique to capture the underlying dynamics of the essential components of the systems allows it to generate links and interactions that lead to a more accurate conclusion and a better understanding of a system. **Figure 3** illustrates a schematic diagram for the different theoretical and quantitative steps involved in system dynamic modelling.

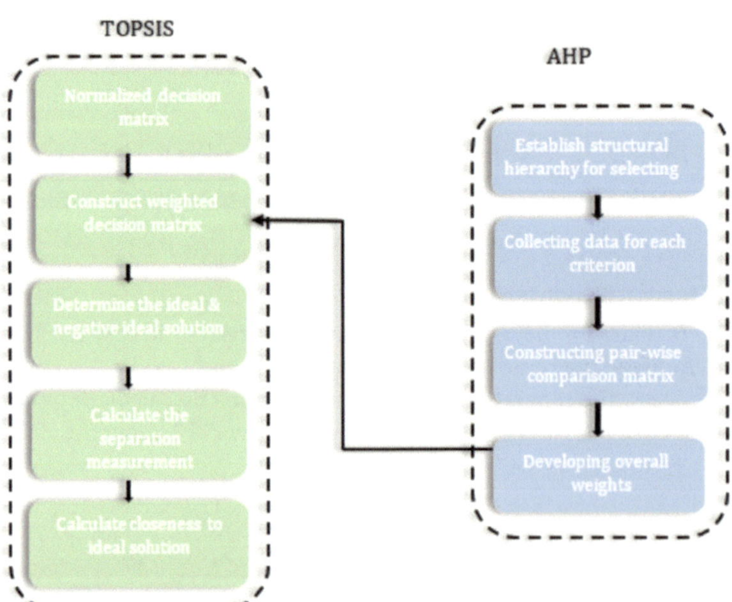

Figure 2.
presents an integrated AHP-TOPSIS for multicriteria decision making.

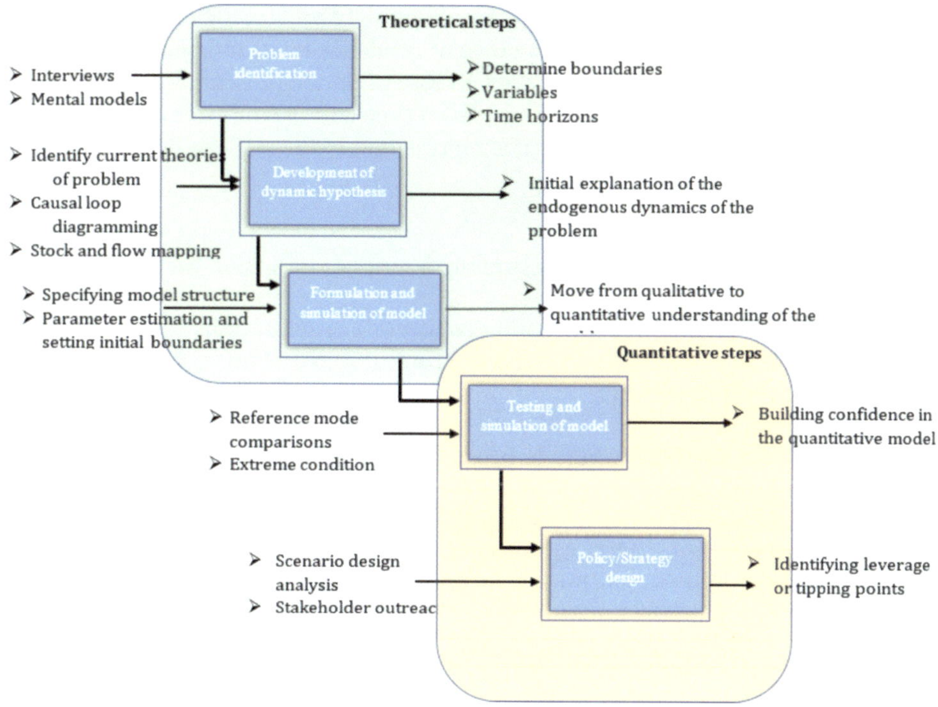

Figure 3.
Schematic diagram showing the steps involved in system dynamic modeling.

3.4 Overview of AR methodology

The AR is generated using two fundamental reactor types (CSTR and PFR) and mixing. The AR construction involves four major steps which include:

i. Generating the PFR trajectory from the feed point.

The PFR Eq. (8) was solved using *MATLAB ode45*, the yield of each extract was plotted as a function of PFR residence time τ using feed concentrations of $C_f = [X_f, P_f, S_f]$ for each extract.

$$\frac{dC}{d\tau} = r(C) \tag{8}$$

$$\mathrm{r(C)} = \begin{bmatrix} \mathrm{r_{yE}} \\ \tau \end{bmatrix} = \begin{bmatrix} \mathrm{f(t,z)} \\ 1 \end{bmatrix} \tag{9}$$

where

$$f(t,z) = \begin{bmatrix} \mu X \\ Y_{PX}\mu X \\ -\frac{1}{Y_{XS}} r_X - \frac{1}{Y_{PX}} r_P - M_s X \end{bmatrix}$$

$$z = (X, P, S)$$

ii. Plot the CSTR locus from the feed point.

The CSTR locus from the feed point C_f is found by solving the non-linear CSTR equation using MATLAB *fsolve* over a range of residence time, and plotting the yield in the $(y_E - \tau)$ space. The data from the CSTR solution are presented as a collection of points and not a line because each residence time corresponds to a different operating scenario. The CSTR relation is represented in the relation below

$$C_f + \tau_1 r(C) - C = 0 \tag{10}$$

iii. Extend the AR boundary by running a series of PFR from each CSTR locus.

Solving the CSTR non-linear equation results in a series of points known as the CSTR locus. These points for each residence time are used as initial feed conditions to generate the PFR trajectory.

iv. Construct the convex hull. In broad reactor network synthesis terms, convex hull can be defined as the smallest subset of a set of points that encloses the original set of points [8]. The convex hull operation was carried out using MATLAB *convhull.* Identifying the convex hull for the set of points helps to identify unique points that can be used for mixing in order to extend the limit of achievability for the system.

3.5 Overview of functional analysis

Figure 4 presents a graphical representation of FAST technique, showing the different phases that will be used in the analysis.

Some of the terms employed in functional analysis are described below

Function: this defines the effect of the produced a product or one of its components to satisfy a need.

Service function: it is the function realized by a product in response to the need of a given user.

Technical function: an internal action of the product defined by the designer within the framework of a solution to assure the service function;

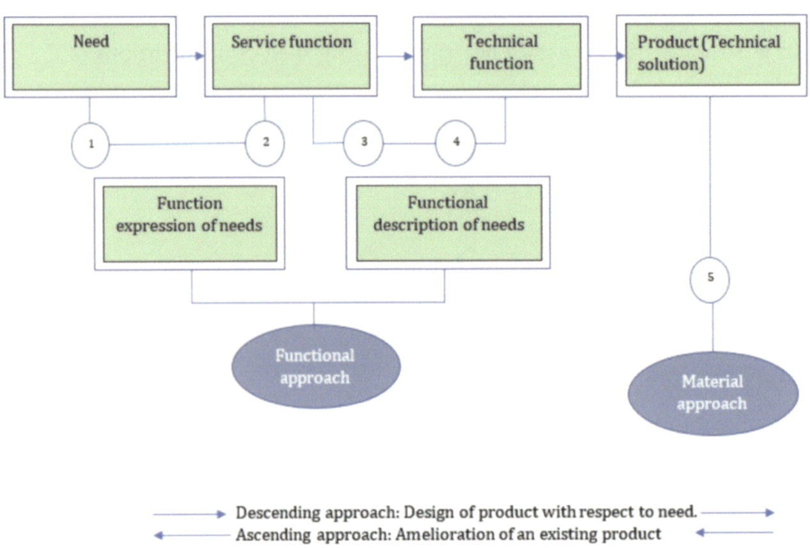

Figure 4.
The different phases of the functional analysis method.

Need: something that is necessary or desired by the customer

The different considerations concerning the functions of the bioenergy technology were obtained by applying the tools of the functional analysis technique.

4. Application case studies

This section deals with how the conceptual framework present in Section 2.1 is applied to a real-time bioenergy project. It's based on the implementation of bioenergy system in Kumasi, Ghana, a Sub-Saharan country in West Africa.

4.1 Deciding on the type of bioenergy system

This stage consists of using MCDM (AHP-TOPSIS) to decide on the type of bioenergy system to install considering the environmental, social, technical, and economic factors of the location.

Table 2 presents the Alternatives Bioenergy Technologies (ABT), which are utilized as potential candidates for installation. The list presented in **Table 2** is not exhaustive but only used to illustrate how the framework can be applied.

The alternative strategies can be evaluated based on multiple attributes, which can be benefit or cost, as shown in **Table 3**. To adhere to the objectives of affordable and clean energy called for by the United Nations Sustainable Development Goal 7, the criteria considered are those that are dominant in determining sustainability of the energy system.

Figure 5 presents a hierarchical decomposition of the decision-making problem summarizing the overall objective, the alternatives, as well as the criteria and sub-criteria used to evaluate the alternatives. This is structured in a well-organized manner such that it shows how each level depends on the upper level.

It is important to note that an initial assumption of equal weights for the major criteria was made, that is economic, environmental, technical, and social factors.

Figure 6 represents the weights of relative importance of each criterion obtained using the AHP method (see steps presented in Section 3.2).

The weights presented in **Figure 6**, implies that safety of the bioenergy technology is more relevant compared to the other criteria. Moreover, the different weights directly reflect the relative importance of environmental impact and safety criteria in the decision making of an alternative energy system.

The next step requires inputting the weights obtained from the AHP into the TOPSIS approach; this results in a ranking of the alternative source of bioenergy.

Symbol	Alternative strategy	Type of process
ABT1	Sugar fermentation to produce bioethanol	Biochemical
ABT2	Anaerobic digestion to produce biogas	Biochemical
ABT3	Transesterification of oils to produce biodiesel	Chemical
ABT4	Biomass gasification to produce syngas	Thermochemical
ABT5	Biomass carbonation to produce biochar	Thermochemical
ABT6	Biomass compression to Briquette	Thermochemical
ABT7	Using microbial fuel cells to generate electricity	Bio-electrochemical

Table 2.
Matrix of bioenergy alternative.

Symbol	Name of criteria	Objective	Description
C1	Efficiency	Maximize	This measured quantity of bioenergy generated per quantity of feed for the different technologies
C2	Safety	Maximized	This measure the treat the technology possess on the employees and environment
C3	Investment cost	Minimize	This measure the capital required to establish the bioenergy technology
C4	Service life	Maximize	This defines how long the technology can sustainable run
C5	Environmental impact (CO_2 emissions)	Minimize	This measures the environmental friendliness of each technology
C6	Land use	Minimize	This describes the land space required to construct each equipment
C7	Job creation	Maximize	This describes the degree of job opportunities generated by each technology
C8	Cost of feedstock	Minimize	This measures the quantity of readily available feedstock and their cost.
C9	Climate dependency	Maximize	Is the strategy optimal for different climatic and/or geographical conditions?

Table 3.
Set of decision criteria to appropriate bioenergy technology selection.

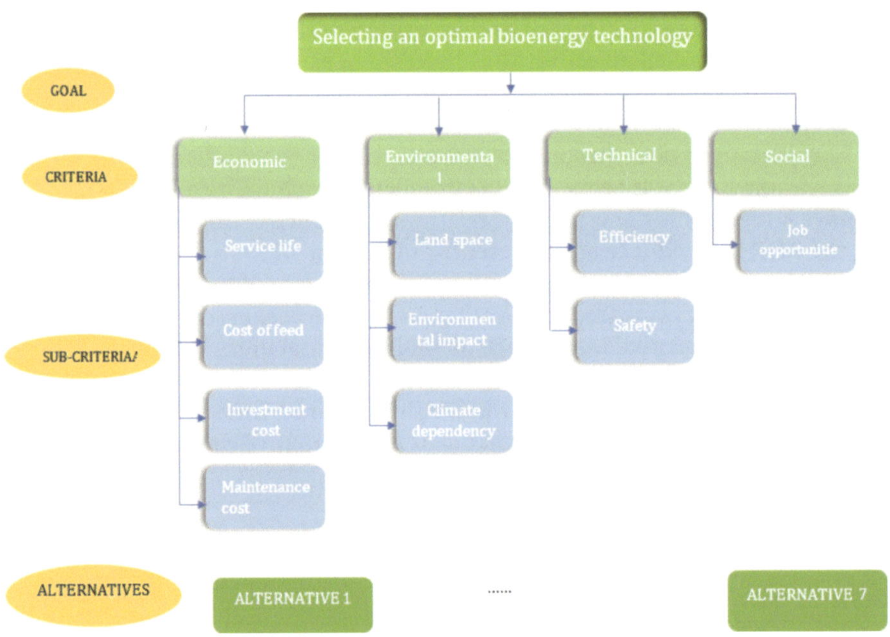

Figure 5.
Hierarchical breakdown of selecting the bioenergy system problem.

Based on the indicators used, lower-ranking of alternatives are more desirable and demonstrate favorability towards sustainability.

From **Figure 7**, sugar fermentation to produce bioethanol is the most appropriate technology for installation in the location of interest. So far, this section has focused on how MCDM tool can be used in selecting an appropriate bioenergy

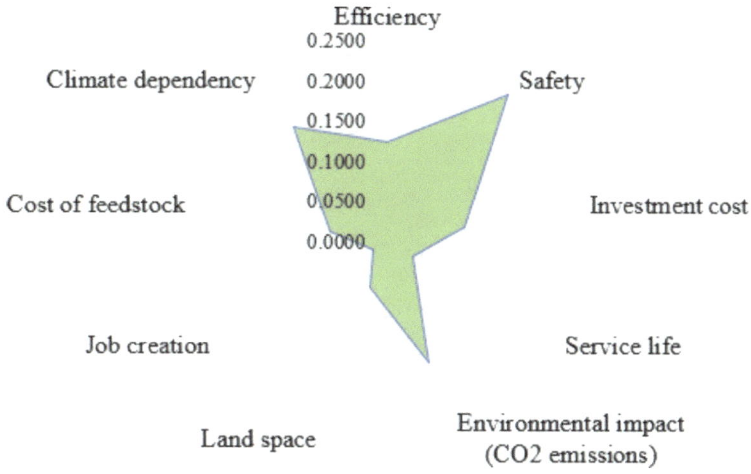

Figure 6.
A spider web diagram that describes the weight of each criterion.

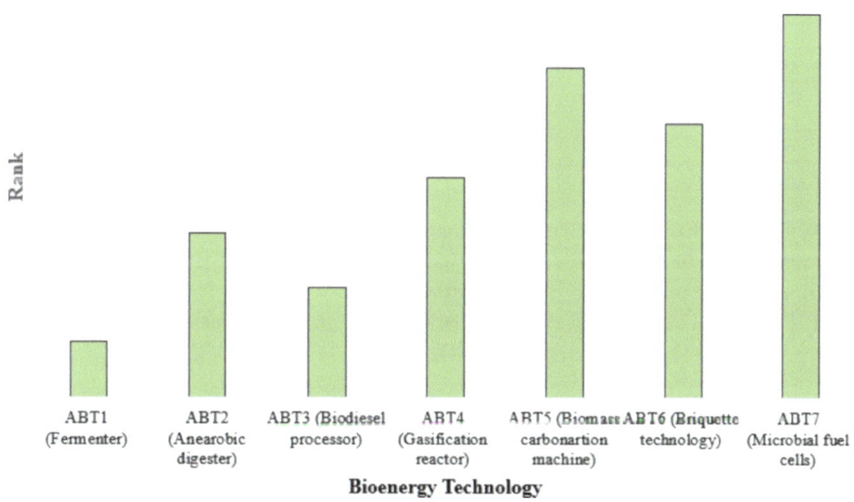

Figure 7.
Presents the ranking of alternative bioenergy technologies after an integrated AHP-TOPSIS approach.

technology. However, before proceeding to the installation of the technology, the next section will discuss how an optimal implementation strategy could be identified using system dynamic modelling.

4.2 Designing an optimal implementation strategy

With the appropriate technology selected, the next step involves the selection of an optimal implementation strategy that requires the development of dynamic models. It is relevant to note that the outcome of the proposed models from the methodology can be used to identify places of management potential (bioenergy policies) and future tipping points that can alleviate potential economic, environmental, and social challenges. The description of the dynamic behavior for bioethanol production was based on the underlying feedbacks and interactions between selected indicators is illustrated through the integrated causal loop diagram in **Figure 1**. The key relevant factors are investment, environmental impact,

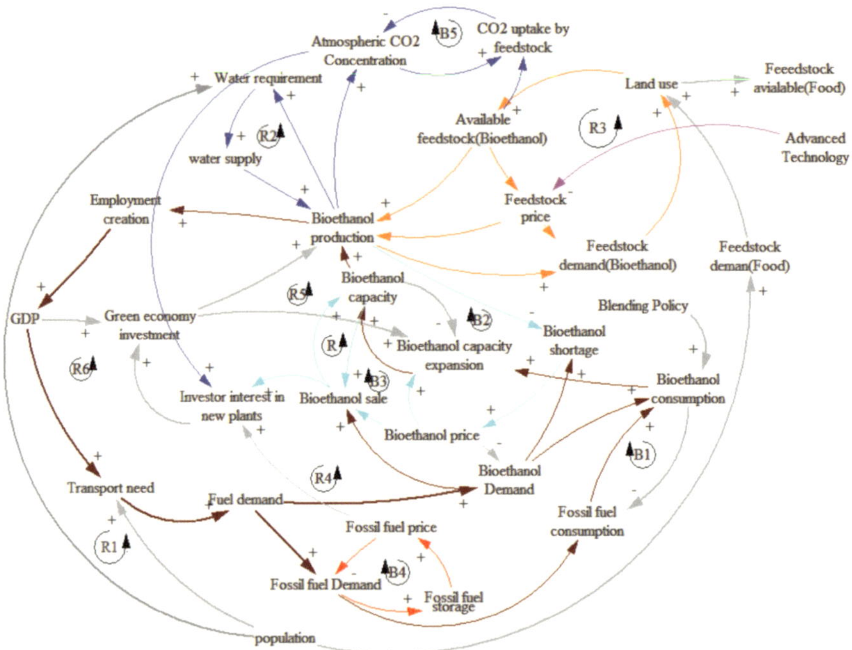

Figure 8.
Casual loop diagram (CLD) for bioethanol production.

employment creation, cost of feedstock, and land space. The causal loop diagrams (CLDs) presented in **Figure 8** are flexible and useful tools for diagramming the feedback structure of systems in any domain [16].

From **Figure 8**, interesting observations can be made considering the reinforcement loop (R) and the balancing loop (B).

1. An increase in population leads to a greater demand for transport (private, commercial, or public), increasing in fuel demand (bioethanol demand) (R1). This will indirectly lead to an increase in fossil fuel and bioethanol demand and consumption. This has the propensity to lead to a shortage in bioethanol, which leads to an increase in prices, but an expansion of bioethanol capacity will lead to an increase in bioethanol production, which in turn will lead to a decrease in bioethanol shortage and price as well. Similarly, an increase in fossil fuel demand will lead to shortage due to production and imports. This can lead to an increase in price and the loop is closed by a decrease in demand due to high prices (loop B4).

2. An increase in bioethanol production leads to an increase in feedstock, which leads to an increase in the land space required (loop (R3)). More importantly, it also leads to the creation of employment. With greater land space being used for feedstock production, price of feedstock will reduce. It is interesting also to notice how an increase in advance method of feedstock production can lead to the use of fewer resources, which directly reduces price of feedstock.

3. An increase in biofuel production leads to an increase in employment and GDP, which again increases the transport need, fuel demand, and biofuels demand and consumption. This loop reverts to increase biofuel production (loop R6). Moreover, an increase in GDP leads to capital available for

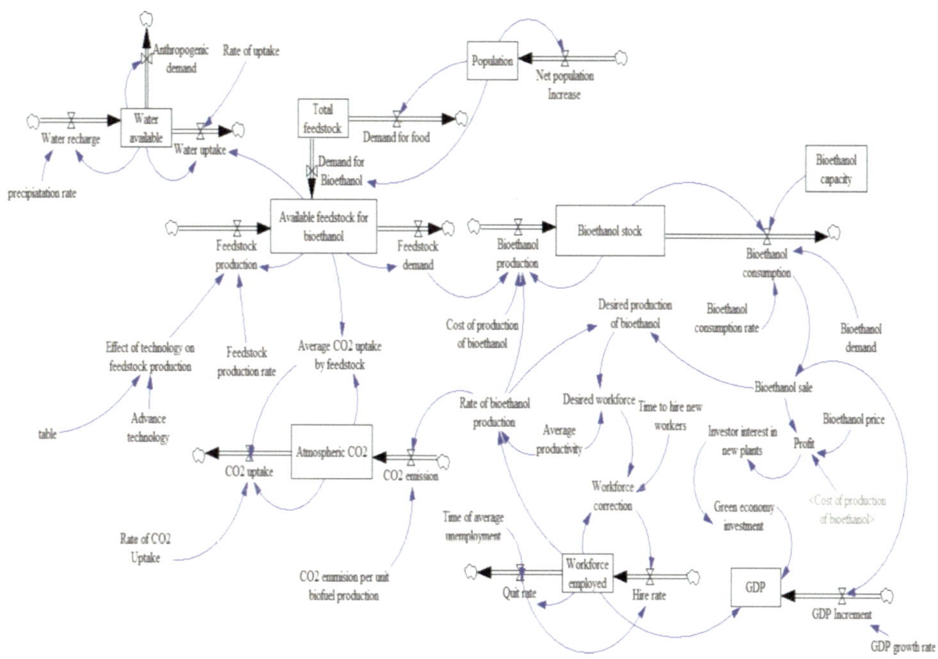

Figure 9.
Stock flow diagram for bioethanol production.

investment into a green economy. This has a direct positive influence on the capacity of bioethanol plants.

While casual loop diagram emphasizes the feedbacks structures within the bioenergy system, stocks and flows diagram amplifies the underlying physical structures of the system. **Figure 9** presents a stock-flow diagram for bioethanol production.

Stocks for this case study include water, bioethanol, feedstock, atmosphere CO_2, populations, workforce, and GDP. These characterize the state of the bioenergy system and generate useful information for policy development. For example, the availability of feedstock for bioethanol has influenced the flow of CO_2 uptake, feedstock production rate, and bioethanol demand. Simulation of these factors was conducted overtime period of 100 months. This indicated a strong correlation between the aforementioned factors and bioethanol production.

Summarily, a hybrid system that works with the national grid is most preferable. This is because such a system will: (1) reduce environmental impact, (2) reduce pressure of land space for feedstock plantation and bioethanol plants, (3) ensure available water for human consumption and (4) most importantly ensure that there is a balance in the quantity of feedstock converted to fuel and consumed as food by the population. Readers need to note that not all the elements of the system were captured, rather key elements that significantly affect the behavior of the system.

4.3 AR construction

Once the optimal implementation strategy had been achieved, the next step is obtaining an optimal fermenter configuration for engineering design and specifications. The technique employed, attainable regions analysis, which is based on the interpretation of the fermentation process as a geometric object by defining a region of achievability that can be attained by the fundamental processes occurring in the fermenter: mixing and bioreaction. The approach captures all possible bioreactor structures and finds the bounds on the performance of the system.

Eqs. (12)–(14) describes the kinetic models that characterize the fermentation of cassava supplemented by malt using *Saccharomyces carlsbergensis*. Cassava extract was selected due to its relative abundance compared to other crops within the region under study. Also, the monod model adopted to capture the substrate, limiting bioreaction taking place, incorporates two-dimensional substrate-product inhibition patterns.

$$\frac{dX}{dt} = r_X = \mu X \tag{11}$$

$$\frac{dP}{dt} = r_P = Y_{PX}\mu X \tag{12}$$

$$\frac{dS}{dt} = r_S = -\frac{1}{Y_{XS}} r_X - \frac{1}{Y_{PX}} r_P - M_s X \tag{13}$$

$$\mu(S,P) = \left(1 - \frac{S}{K_{is}}\right)\frac{\mu_{max}S}{K_{sx} + S}\left(1 - \frac{P}{K_{iP}}\right) \tag{14}$$

Eq. (15) is substituted into the dynamic relations (Eqs. (12)–(14)).

Before constructing the AR, it is expedient to determine the dimension in which the AR will reside. The dimensions of the AR depend on the number of independent reactions taking place. From Eq. (5), only one independent reaction involving three components (X–S–P) is taking place; hence the AR constructed must reside in a one-dimensional subspace of $\mathbb{R}^3$ for all achievable set of points.

$$S_1 + S_2 + \ldots + S_n \xrightarrow{X} P_1 + P_2 + \ldots + P_m \tag{15}$$

Since the dimension of the AR is one, we, therefore, need to select a variable which contains the effect of all three states for which case the bioethanol yield (y_p) has been selected. To enable graphical visualization of the AR, another variable, residence time (τ_p) will be added to the system such that it can be plotted in a two-dimensional space. A major consideration when selecting variables for plotting attainable regions is that the variables must follow the linear mixing law. It has been reported in literature by Mings et al. and Abunde et al., how the residence time and yield follow the linear mixing law [8, 20, 22]. This is represented by Eqs. (17) and (18).

$$\tau^* = \lambda\tau_1 + (1-\lambda)\tau_2 \tag{16}$$

$$y_E^* = \lambda y_E^1 + (1-\lambda)y_E^2 \tag{17}$$

Where τ_1 and τ_2 are the residence time in two reactors and τ^* is the residence time upon mixing. y_E^1 and y_E^2 are the yields of reactor 1 and reactor 2, respectively, and y_E^* is the yield upon mixing.

With the kinetic model and initial conditions now known, we begin constructing the AR by generating the PFR trajectory and then the CSTR locus, then generating PFR trajectories using the CSTR locus, as illustrated in **Figure 10**.

From **Figure 10**, the boundary of the candidate AR can be defined by two main reactor configurations: (1) A CSTR followed by a PFR and (2) A CSTR followed by a PFR with a bypass from the feed to the effluent stream. This implies that all the ethanol yield contained within the defined region can be achieved by the above reactor types, with differences coming at the level of the residence time. Furthermore, the operating limits of the system (defined by the area of the convex hull) are

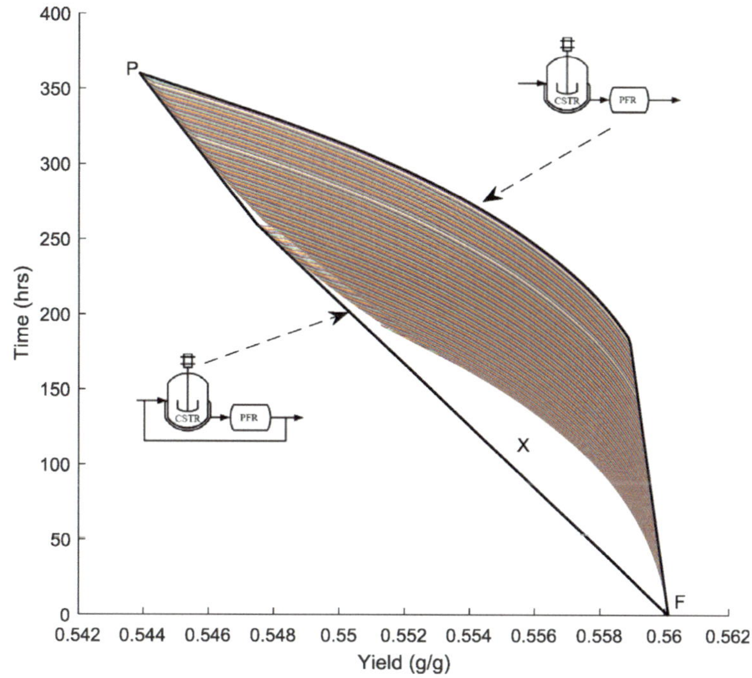

Figure 10.
Two-dimensional candidate attainable regions for Cassava extract using two-dimensional sudden stop substrate and product inhibition patterns.

1.762 (g/g hrs), which provides a geometric representation of all possible yields that can be achieved by the aforementioned reactor structures. It can also be inferred from the figure that using a fermenter structure (A CSTR followed by a PFR with a bypass) as oppose to a single fermenter reduces the overall residence time of the fermentation process. More interestingly, yields within the region of X, which were initially not achievable, are now achievable by using a fermenter structure.

Once the candidate AR has been constructed for a given kinetic and feed concentration, the limits of achievability of the system are defined. The candidate AR generated can, therefore, be used to answer design questions and determine performance targets by developing appropriate objective functions, which can be overlaid as contours on the AR boundaries. For illustration, an economic index such Payback period, which is key to investors, is considered. Economic models were then developed, incorporating the dimension of the AR and overlaid as contours on the candidate AR.

Figure 11 illustrates the payback periods overlaid on the Candidate AR to obtain optimal operating points and corresponding reactor structures by identifying the points of intersection of the objective function on the AR boundary.

Interestingly, two major observations can be made from **Figure 10**: (1) The range of payback periods considered intersect the AR at many points in the region, indicating that there are multiple operating points (multiple optima) for this system. Therefore the actual operating points to be selected vehemently depend on other auxiliary factors such as the investor's available capital. (2) Shorter PBP are achievable for higher yields at lower reactor volumes. This is interesting because an investment that involves smaller reactor volumes (lower investment cost) and higher operating yields (higher annual benefits) should require a shorter to recover investment. (3) Another interesting observation is that, as the payback period increase, the influence of running cost (reactor volume) on the PBP decreases.

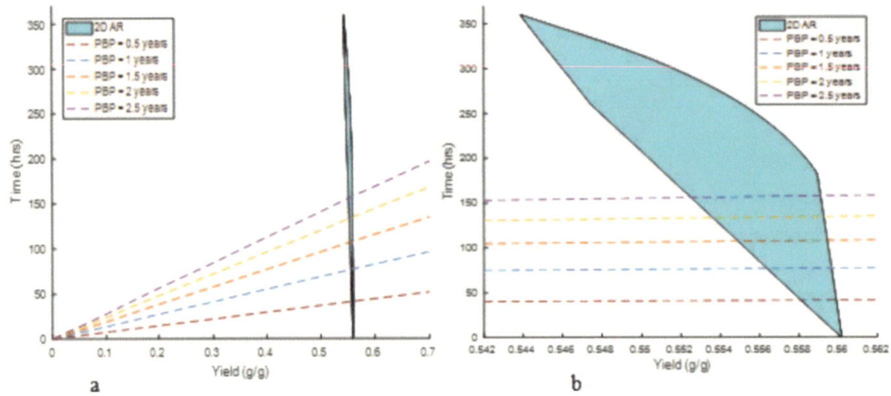

Figure 11.
Different contours of payback period overlaid onto the AR for cassava extract to determine the optimal operational points ((b) is a closer zoom of (a), demonstrating how the contours of the PBP intersect the AR).

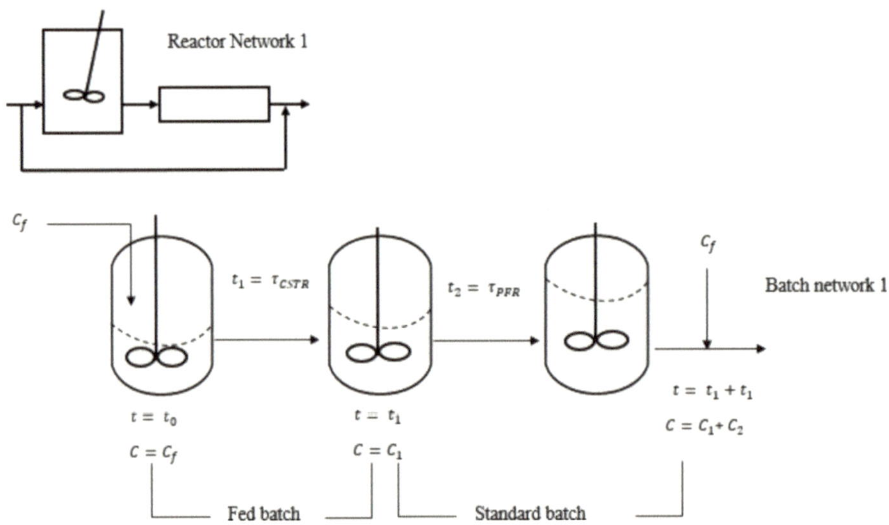

Figure 12.
Optimal continuous fermenter structure and its corresponding batch fermenters.

This is observed from **Figure 11a** by the closeness of the contours from 1.5, 2, and 2.5 years. Therefore, it is sensible to construct a fermenter volume that is larger for payback periods between 1.5 and 2.5 years, since the cost influence is minimal, and that reactor volume can be used to achieve all desired payback periods.

In summary, the AR theory presents a geometric technique that can be used to identify optimal process configuration. Therefore **Figure 12** illustrates the optimal continuous reactor and its corresponding batch fermenter for bioethanol production.

5. FAST analysis

5.1 Functional analysis of need

The Horned Beast diagram, illustrated in **Figure 13**, provides a visual tool that seeks to answer three fundamental questions:

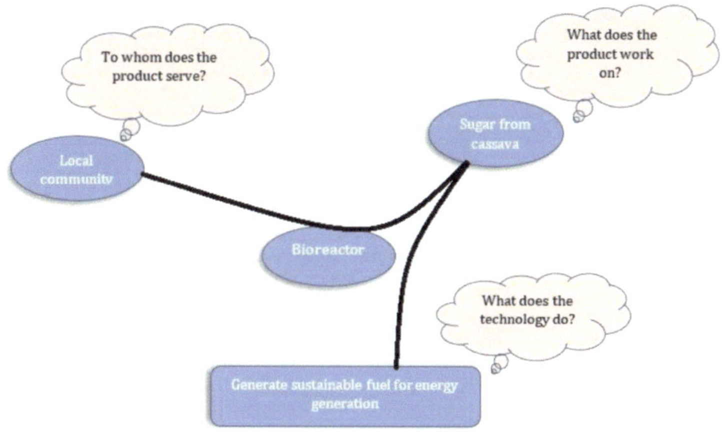

Figure 13.
The horned beast diagram above is used to determine the needs to which the technology answers.

1. Who will the product serve?

2. What does the product interact with?

3. What does the product do?

5.2 The octopus diagram

Figure 14 presents an Octopus diagram that comprises the product in question to be designed and the different components of its external medium. The figure further describes the elements associated with the bioreactor and its environment.

The above functions involved in **Figure 14** are elucidated in **Table 4**. The advantage of the octopus diagram is that it helps to visualize and validate the elements of the technology.

Figure 15 shows a FAST diagram that presents the technological solutions which permit the satisfaction of the principal and constraint functions.

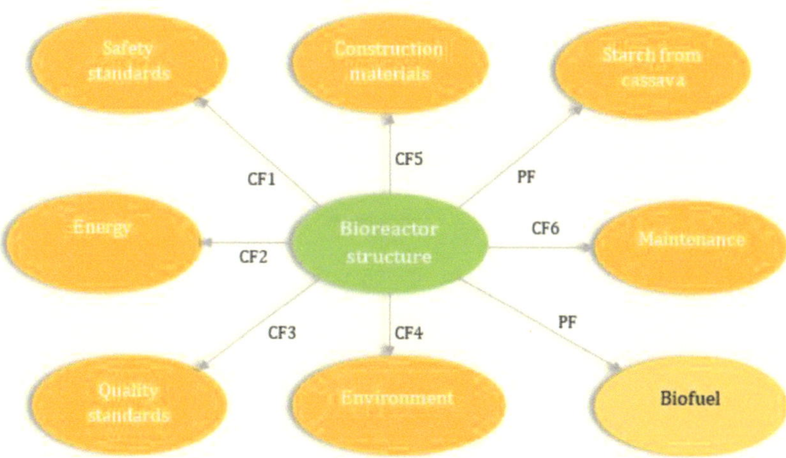

Figure 14.
The octopus diagram showing the relationship between the bioenergy system and its environment.

	Principal function
PF	Conversion of starch to bioethanol in order to generate energy
	Constraint functions
CF1	The biofuel should meet all required safety standards and minimize losses from accident
CF2	The technology should use a renewable energy source
CF3	The biofuel produced should meet international standards for fuels
CF4	The technology should have a less environmental impact
CF5	Material of construction should be available and less expensive
CF6	Maintenance should be simple and easily carried out routinely

Table 4.
Principal functions together with the constraints.

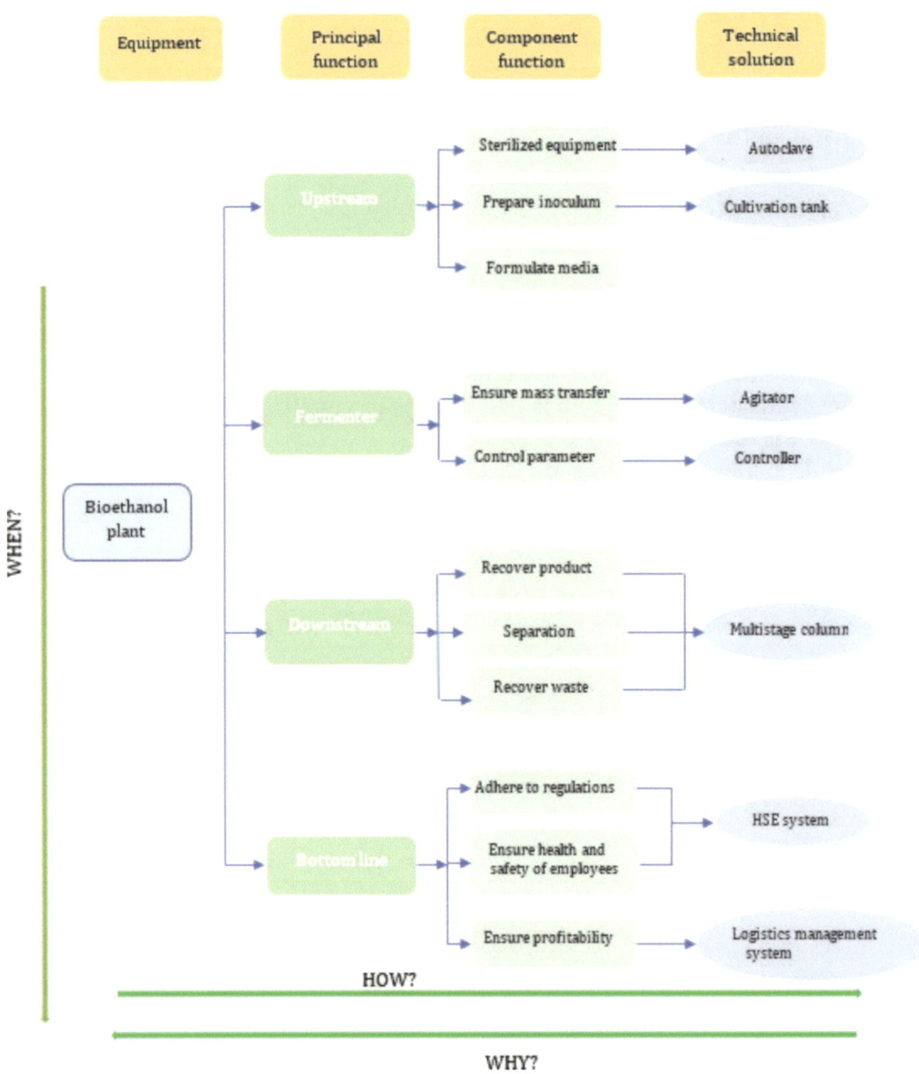

Figure 15.
FAST diagram showing functions and their corresponding technical solutions.

Once the FAST diagram is constructed, the next step is to develop a Value Analysis Matrix that examines the component costs of the equipment in relation to the value perceived by the customer. Value Analysis Matrix, also known as Functional-Cost Matrix, was derived from the Quality Functional Deployment Methodology. The strength of this technique is that it associates the functions of a product back to the customer's needs. It can also develop mechanisms that relate to functions as either strongly, moderately, or weakly supporting the given function and can also be used to calculate each mechanism's relative weight in satisfying the designated functions. This enables management to check whether the money spent on function and component is worth it. For illustrations, the approach was not exhausted in this work. Once exhausted, management can move further into the equipment specification, then the installation of the equipment.

6. Conclusion

Returning to the challenges posed at the beginning of this chapter, it is now possible to state that (1) MCDM provides an appropriate technique for selecting and assessing an optimal bioenergy technology (bioethanol fermenter) that seeks to address the social, economic, technical and environmental factors for sustainable development. (2) A hybrid energy system that comprises of the bioethanol plant along with the national grid proved at optimal implementation strategy as it ensured a balance in the bioenergy system. It is quite interesting to notice how system dynamics modeling presents an efficient tool to model and simulate energy systems and their interaction with other systems, as demonstrated in this chapter. The tool was used to investigate the economic, environmental and social impact of bioethanol production in view of respecting sustainability criteria while striking a balance between the several subsystems involved (3) the optimal fermenter structure required for the fermentation of the different extracts includes a CSTR followed by a PFR as well as a CSTR followed by a PFR with bypass from feed. And finally, a value analysis was conducted to identify the components required for the technology to meet its functions. More importantly, the methodological framework presented an exciting and thrilling route to how sustainable technologies could be successfully installed. This chapter has gone some way towards enhancing our understanding of how model-based approaches relative to conventional implementation strategies ensure sustainable development. A model-based approach to delivering sustainable solutions is gradually becoming an exhilarating area for sustainable systems engineers. Readers should expect electrifying exploits from the authors as they seek to leverage on model-based techniques, Artificial Intelligence (AI), and digital technology to unlock Africa's potential in the food-water-energy-health nexus.

Acknowledgements

Our team expresses gratitude to the following institutions; The Brew-Hammond Energy Centre, KNUST Ghana, The Water and Environmental Engineering Group, NTNU Ålesund and the Abunde Sustainable Engineering Group (AbundeSEG) for its immense technical support.

Conflict of interest

The authors declare that they have no known competing financial interests or personal relationships that could have appeared to influence the work reported in this paper.

Funding

This work was supported by EnPe - NORAD under the project Upgrading Education and Research Capacity in Renewable Energy Technologies (UPERC-RET).

Nomenclature

$\left(r_{ij}\right)_{m\times n}$	normalized matrix
λ_{max}	maximum Eigen value
n	number of attributes
A	pairwise comparison matrix
W	the estimate of the decision-makers weight
τ_1	residence time

Author details

Fabrice Abunde Neba[1,2*], Prince Agyemang[1], Yahaya D. Ndam[1], Endene Emmanuel[1], Eyong G. Ndip[1] and Razak Seidu[2]

1 Abunde Sustainable Engineering Group (AbundeSEG), Buea, Cameroon

2 Institute of Marine Operations and Civil Engineering, Norwegian University of Science and Technology, Ålesund, Norway

*Address all correspondence to: fabrice.a.neba@ntnu.no

References

[1] Chatterjee R, Gajjela S, Thirumdasu RK. Recycling of organic wastes for sustainable soil health and crop growth. International Journal of Waste Resources. 2017;**07**(03). DOI: 10.4172/2252-5211.1000296

[2] Smith A, Brown K, Ogilvie S, Rushton K, Bates J. Waste Management Options and Climate Change: Final Report to the European Commission. DG Environment. 2001

[3] Gingerich DB, Mauter MS. Air Emission Reduction Benefits of Biogas Electricity Generation at Municipal Wastewater Treatment Plants. 2018

[4] Koch K, Helmreich B, Drewes JE. Co-digestion of food waste in municipal wastewater treatment plants: Effect of different mixtures on methane yield and hydrolysis rate constant. Applied Energy. 2015;**137**:250-255

[5] PNUMA. UN Environment Annual Report, Empowering People to Protect the Planet. 2016, p. 20

[6] Chel A, Kaushik G. Renewable energy technologies for sustainable development of energy efficient building. Alexandria Engineering Journal. 2018;**57**(2):655-669

[7] Mustafa A et al. Renewable energy technologies and charaterization. **1**(4) (TR-109496):102-116

[8] Ming D, Glasser D, Hildebrandt D, Glasser B, Metzer M. Attainable Region Theory: An Introduction to Choosing an Optimal Reactor. Hoboken, New Jersey: John Wiley & Sons, Inc; 2016

[9] Ming D, Glasser D, Hildebrandt D. Application of attainable region theory to batch reactors. Chemical Engineering Science. 2013;**99**:203-214

[10] Hildebrandt D. Synthesis of chemical reactor networks. 1995

[11] Snegirev DA, Valiev RT, Eroshenko SA, Khalyasmaa AI. Functional assessment system of solar power plant energy production. In: Proc. 8th Int. Conf. Energy Environ. Energy Saved Today is Asset Futur. CIEM 2017. 2017. pp. 349-353

[12] Wüstenhagen R, Wolsink M, Bürer MJ. Social acceptance of renewable energy innovation: An introduction to the concept. Energy Policy. 2007;**35**(5):2683-2691

[13] Wang JJ, Jing YY, Zhang CF, Zhao JH. Review on multi-criteria decision analysis aid in sustainable energy decision-making. Renewable and Sustainable Energy Reviews. 2009; **13**(9):2263-2278

[14] Ibáñez-Forés V, Bovea MD, Pérez-Belis V. A holistic review of applied methodologies for assessing and selecting the optimal technological alternative from a sustainability perspective. Journal of Cleaner Production. 2014;**70**:259-281

[15] Mardani A, Jusoh A, Nor KMD, Khalifah Z, Zakwan N, Valipour A. Multiple criteria decision-making techniques and their applications—A review of the literature from 2000 to 2014. Economic Research-Ekonomska Istraživanja. 2015;**28**(1):516-571

[16] Sterman J. Business dynamics: Systems thinking and modeling for a complex world. Boston: Irwin/McGraw-Hill; 2000

[17] Turner BL, Menendez HM, Gates R, Tedeschi LO, Atzori AS. System dynamics modeling for agricultural and natural resource management issues: Review of some past cases and

forecasting future roles. Resources. 2016;**5**(4)

[18] Asiedu N, Hildebrandt D, Glasser D. Experimental simulation of three-dimensional attainable region for the synthesis of exothermic reversible reaction: Ethyl acetate synthesis case study. Industrial & Engineering Chemistry Research. 2015;**54**(10): 2619-2626

[19] Abunde Neba F, Jiokap Nono Y. Modeling and simulated design: A novel model and software of a solar-biomass hybrid dryer. Computers & Chemical Engineering. 2017;**104**:128-140

[20] Neba FA, Asiedu NY, Addo A, Seidu R. Attainable regions and fuzzy multi-criteria decisions: Modeling a novel configuration of methane bioreactor using experimental limits of operation. Bioresource Technology. 2019:122273

[21] Neba FA, Tornyeviadzi HM, Østerhus SW, Seidu R. Self-optimizing attainable regions of the anaerobic treatment process: Modeling performance targets under kinetic uncertainty. Water Research. 2020: 115377

[22] Abunde Neba F, Asiedu NY, Addo A, Morken J, Østerhus SW, Seidu R. Simulation of two-dimensional attainable regions and its application to model digester structures for maximum stability of anaerobic treatment process. Water Research. 2019;**163**:114891

[23] Asiedu NY, Hildebrandt D, Glasser D. Batch distillation targets for minimum energy consumption. Industrial & Engineering Chemistry Research. 2014;**53**(7):2751-2757. DOI: 1021/ie402044y

[24] Asiedu N, Hildebrandt D, Glasser D. Experimental simulation of a two-dimensional attainable region and its application in the optimization of production rate and process time of an adiabatic batch reactor. Industrial & Engineering Chemistry Research. 2014; **53**(34):13308-13319

[25] Metzger MJ, Glasser D, Hausberger B, Hildebrandt D, Glasser BJ. Use of the attainable region analysis to optimize particle breakage in a ball mill. Chemical Engineering Science. 2009;**64**(17):3766-3777

[26] Mota P, Campos AR, Neves-Silva R. First look at MCDM: Choosing a decision method. Adv. Smart Syst. Res. 2013;**3**(2):25-30

[27] Wątróbski J, Jankowski J, Ziemba P, Karczmarczyk A, Zioło M. Generalised framework for multi-criteria method selection. Omega (United Kingdom). 2019;**86**:107-124

[28] Akash BA, Mamlook R, Mohsen MS. Multi-criteria selection of electric power plants using analytical hierarchy process. Electric Power Systems Research. 1999;**52**(1):29-35

[29] Jaber JO, Jaber QM, Sawalha SA, Mohsen MS. Evaluation of conventional and renewable energy sources for space heating in the household sector. 2008; **12**:278-289

[30] Saaty RW. The analytic hierarchy process-what it is and how it is used. Mathematical Modelling. 1987;**9**(3–5): 161-176

Chapter 6

Energy Potential of Biomass Sources in Slovakia

Andrea Majlingová, Martin Lieskovský, Maroš Sedliak and Marián Slamka

Abstract

Renewable energy has provided many potential benefits, including a reduction in greenhouse gas (GHG) emissions, the diversification of energy supplies, and a reduced dependency on fossil fuel markets (oil and gas in particular). The growth of renewable energy sources (RES) may also have the potential to stimulate employment in the European Union (EU), through the creation of jobs in new green technologies. In this chapter, first, we introduce the information on renewable energy sources, their statistics, and legislation background in Slovakia. In more detail, we further introduce the information on forest and agricultural biomass as a renewable energy source. In the experimental part, we introduce two case studies—the assessment of the potential stock of woody biomass and the determination of energetic properties of woody biomass, i.e., selected fast-growing tree species based on the implementation of laboratory fire tests and calorimetric analyses.

Keywords: woody biomass, energy potential, stock, renewable energy source

1. General overview on renewable energy production in EU and in Slovakia

In general, renewable energy sources (RES) include wind power, solar power (thermal, photovoltaic, and concentrated), hydropower, tidal power, geothermal energy, ambient heat captured by heat pumps, biofuels, and the renewable part of waste.

Here we introduce the overview of statistics on renewable energy sources in the EU published by Eurostat [1].

The information presented here is based on data compiled in accordance with accounting rules set down in the Directive 2009/28/EC [2] on the promotion of the use of energy from renewable sources and calculated on the basis of energy statistics covered by Regulation 1099/2008 on energy statistics, most recently amended in November 2017 by Commission Regulation 2017/2010. The most recent data available on the share of energy from renewable sources are for the reference year 2017.

The primary production of renewable energy within the EU-28 in 2017 was 226.5 million tons of oil equivalent (toe). The quantity of renewable energy produced within the EU-28 increased overall by 64.0% between 2007 and 2017, equivalent to an average increase of 5.1% per year [1].

Among renewable energies, the most important source in the EU-28 was wood and other solid biofuels, accounting for 42.0% of primary renewable production in 2017 (**Figure 1**).

Wind power was, for the first time, the second most important contributor to the renewable energy mix (13.8% of the total), followed by hydropower (11.4%). Although their levels of production remained relatively low, there was a particularly rapid expansion in the output of biogas, liquid biofuels, and solar energy, which accounted, respectively, for a 7.4, 6.7, and 6.4% share of the EU-28's renewable energy produced in 2017. Ambient heat (captured by heat pumps) and geothermal energy accounted for 5.0 and 3.0% of the total, respectively, while renewable wastes increased to reach 4.4%. There are currently very low levels of tide, wave, and ocean energy production, with these technologies principally found in France and the United Kingdom [1].

In 2018, the share of energy from renewable sources in gross final energy consumption reached 18.0% in the European Union (EU), up from 17.5% in 2017 and more than double the share in 2004 (8.5%), the first year for which the data are available.

Gross final consumption of energy is defined in the Renewable Energy Directive 2009/28/EC [2] as the energy commodities delivered for energy purposes to industry, transport, households, services (including public services), agriculture, forestry, and fisheries, including the consumption of electricity and heat by the energy branch for electricity and heat production and including losses of electricity and heat in distribution and transmission [1].

The increase in the share of renewables is essential to reach the EU climate and energy goals. The EU's target is to reach 20% of its energy from renewable sources by 2020 and at least 32% by 2030.

The European Council endorsed a 2030 Framework for Energy and Climate for the Union based on four key Union-level targets: a reduction of at least 40% in economy-wide greenhouse gas (GHG) emissions; an indicative target of improvement in energy efficiency of at least 27%, to be reviewed by 2020 with a view to increasing the level to 30%; a share of renewable energy consumed in the Union of at least 27%; and electricity interconnection of at least 15%. It specified that the target for renewable energy is binding at Union level and that it will be fulfilled through Member States' contributions guided by the need to deliver collectively the Union target [3].

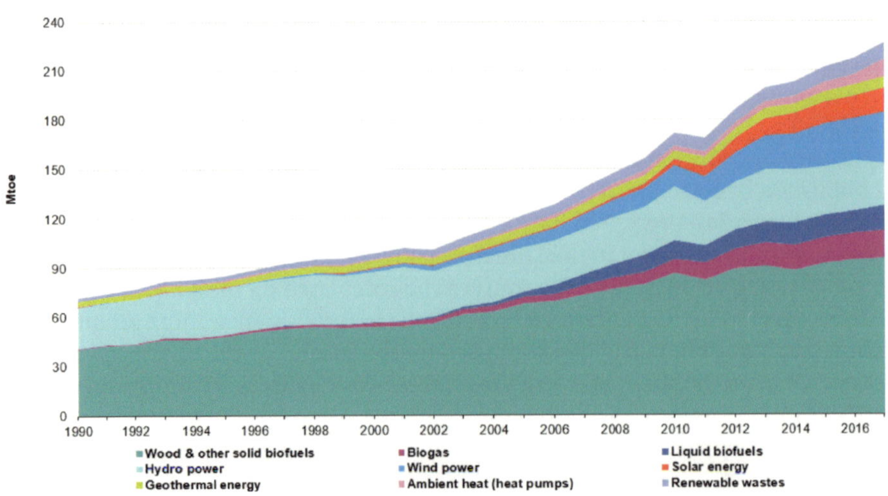

Figure 1.
Primary production of energy from renewable sources in EU-28 in period 1990–2017 (Source: Eurostat [1]).

A recast of Directive 2009/28/EC [2] of the European Parliament and of the Council has introduced a new, binding, renewable energy target for the Union for 2030 of at least 32%, including a provision for a review with a view to increasing the Union-level target by 2023. Amendments to Directive 2012/27/EU [4] of the European Parliament and of the Council have set the Union-level target for improvements in energy efficiency in 2030 to at least 32.5%, including a provision for a review with a view to increasing the Union-level targets.

This target is distributed between the EU Member States with national action plans designed to plot a pathway for the development of renewable energies in each of the Member States [5].

Among the 28 EU Member States, 12 Member States have already reached a share equal to or above their national 2020 binding targets: Bulgaria, Czechia, Denmark, Estonia, Greece, Croatia, Italy, Latvia, Lithuania, Cyprus, Finland, and Sweden. Four Member States are close to meet their targets (i.e., less than 1 percentage point (pp) away), nine are between 1 and 4 pp. away, while three are 4 or more pp. away from their targets [1].

The share of renewable energy in gross final energy consumption is identified as a key indicator for measuring progress under the Europe 2020 strategy for smart, sustainable, and inclusive growth. This indicator may be considered as an estimate for the purpose of monitoring Directive 2009/28/EC [2] on the promotion of the use of energy from renewable sources—however, the statistical system in some countries for specific renewable energy technologies is not yet fully developed to meet the requirements of this Directive.

Figure 2 shows the latest data available for the share of renewable energies in gross final energy consumption and the targets that have been set for 2020. The share of renewables in gross final energy consumption stood at 18% in the EU-28 in 2018, compared with 8.5% in 2004.

This positive development has been prompted by the legally binding targets for increasing the share of energy from renewable sources enacted by Directive 2009/28/EC [2] on the promotion of the use of energy from renewable sources.

The share of energy from renewable sources is divided into three different components: share in electricity, share in heating and cooling, and share in transport.

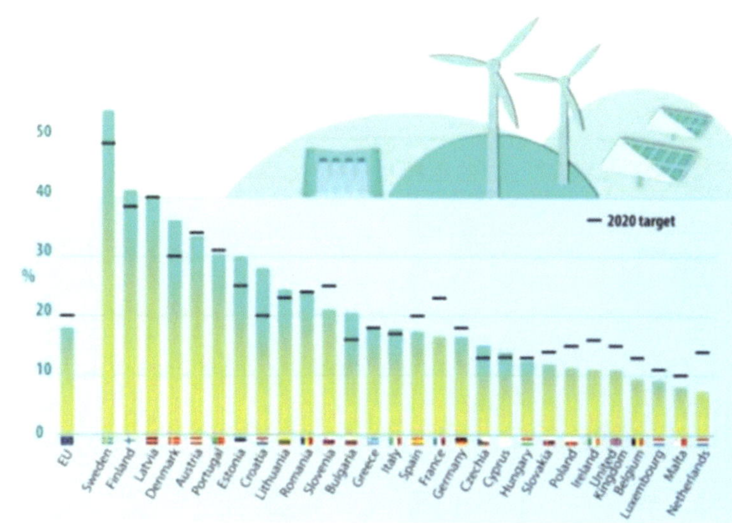

Figure 2.
Share of energy from renewable sources in in EU-28 in % of gross final energy consumption in 2018 (Source: Eurostat [1]).

While the EU as a whole is on course to meet its 2020 targets, some Member States will need to make additional efforts to meet their obligations as regards the two main targets: the overall share of energy from renewable sources in the gross final energy consumption and the specific share of energy from renewable sources in transport [1].

In 2017, electricity generation from renewable sources contributed more than one quarter (30.7%) to total EU-28 gross electricity consumption. Wind power was for the first time the most important source, followed closely by hydropower.

Renewable energy accounted for 19.5% of total energy used for heating and cooling in 2017. This was a significant increase from 10.4% in 2004. Increases in industrial sectors, services, and households (building sector) contributed to this growth [1].

But the Slovak Republic (SR) is moving away from its target for the share of renewable energy sources. This is set at 14% for 2020.

In 2017, however, Slovakia reached only 11.5%, while the share decreased for the second consecutive year. In 2016, it was 12%. In 2015, it was 12.9%. Slovakia returned statistically before 2014, when the share was 11.7% [1].

The share of energy from renewable sources in final energy consumption in the Slovak Republic in period 2004–2016 is shown in **Figure 3**.

The decrease in the share of renewable energy sources was caused by lower growth in the use of renewable energy sources than the growth in final energy consumption. The growth in electricity consumption and the significant increase in the use of motor fuels, which caused a dynamic increase in energy consumption, reflect the Slovak Republic's economic growth. In the long term, the Slovak Republic's priority is energy efficiency, which leads to a reduction in energy consumption and thus to savings in fossil fuels and greenhouse gas emissions.

At the same time, in 2017, the highest increase in energy consumption in Slovakia was recorded by 7% of all EU Member States. Slovak gross domestic product (GDP) increased this year by 3.2%. This means that the country is failing to separate energy consumption from economic growth and thus enhance energy efficiency.

Of all 28 EU Member States, in the share of renewable energy, the Slovak Republic ended in the ninth place backward.

Figure 3.
Share of energy from renewable sources in final energy consumption in SR in period 2004–2016 (Source: Energie Portál [6]).

Year	Natural gas (GWh)	Biomass (Kt)	Coal (Kt)	Biogas (GWh)	Fuel oil (Kt)
2016	8514	1113	571	275	96
2017	8141	845	577	326	128
2018	8637	877	586	326	128

Table 1.
Renewable energy sources are used in addition to electricity production also for heat production (Source: URSO [7, 8]).

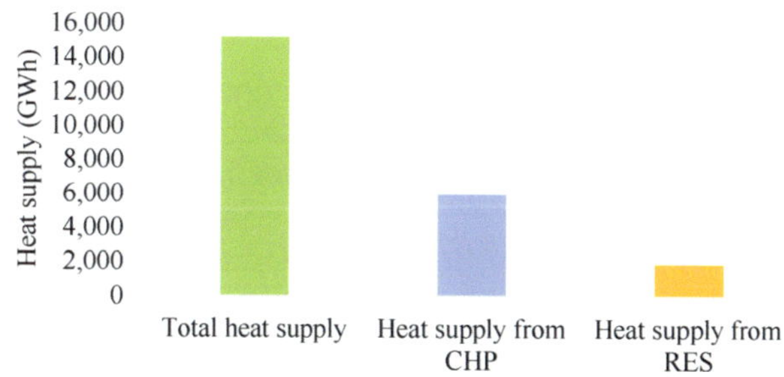

Figure 4.
Heat supply in SR in 2018 (Source: URSO [7, 8]).

According to the latest statistics of Eurostat [1], the Slovak Republic is the country with the highest year-on-year increase in final energy consumption—by 7%.

As energy consumption in the Slovak Republic is growing and renewable sources are not developing, their share inevitably decreases.

In the Slovak Republic, electricity from renewable sources is promoted through a fixed feed-in tariff. Energy companies are obliged to purchase and pay for electricity exported to the grid.

Renewable energy biomass must be given priority connection, and electricity from renewable sources must be given priority dispatch. The grid operator is obliged to extend the grid without discriminating against certain users.

Renewable energy sources are used in addition to electricity production also for heat production. URSO [7, 8] 2019 statistics show that in 2018 the most used fuel for heat production in Slovakia was natural gas. As can be seen from the data in **Table 1**, its use increased year-on-year most significantly from all fuels to around 8640 GWh.

The total volume of heat supply from renewable energy sources in 2018 was less than 2000 GWh according to data from the URSO Annual Report [7, 8]. From the combined heat and power (CHP) came 6000 GWh. The distribution of heat supply volume is visualized in **Figure 4**.

The support of heat from renewable energy sources mainly takes the form of financial support for investments in the Slovak Republic.

Further, we are focusing more on the legislation governing the use of renewable energy sources in Slovakia.

2. Legislation governing the use of renewable energy sources in Slovakia

In Slovakia, the primary legislation consists of Act of the National Council of the Slovak Republic No. 656/2004 Coll. Energy Act [9]. This Act defines the basic processes related to electricity and RES, as well as basic concepts, and performance of state administration. It also introduces conditions for issuing a license for electricity production from RES and conditions for the construction of energy facilities (including facilities for electricity production, when electricity is produced from RES). And also, the Act defines the rights and obligations of a producer of electricity from RES and rights and obligations of the transmission and distribution system operator to which the producer of the electricity from RES is connected and through which the transmission or distribution of electricity produced from RES is carried out to the final consumption point. Under this Act, support for RES is achieved through the priority access, connection, transmission, distribution, and supply of electricity produced from RES. However, the producer must respect the technical and commercial conditions of access and grid connection, which are specified in the tertiary legislation.

The secondary legislation consists of the Government Regulation No. 211/2010 [10], laying down the rules for the functioning of the electricity market and the Act no. 309/2009 Coll. [11] on the Promotion of Renewable Energy Sources and High Efficiency Cogeneration and on Amendments to Certain Acts.

Government Regulation No. 211/2010 [10] (Electricity Market Rules), by its very nature, supplements the "Energy Act" and specifies some of its provisions. These market rules lay down the conditions for connection, access, transmission, and distribution of electricity. It defines the criteria to connect the producer to the system, criteria to carry out the distribution of electricity produced also from RES, and the necessary contractual relations necessary for connecting the production equipment. The contractual relations named in the Market Rules are further specified in tertiary legislation. The electricity market rules further define and develop functional processes related to market participant deviation, registration of daily supply diagrams, etc.

The Act no. 309/2009 Coll. [11] specifies the method of support and conditions for the promotion of electricity production from renewable energy sources, electricity by high-efficiency cogeneration, and biomethane; rights and obligations of producers of electricity from renewable energy sources, electricity from cogeneration, electricity from high-efficiency cogeneration, and biomethane; the rights and obligations of other electricity and gas market participants; and the rights and obligations of the legal person or the natural person who places on the market fuels and other energy products used for transport purposes.

The tertiary legislation includes in particular rules of operation of the transmission system operator; operating rules of the distribution system operator; technical conditions of the transmission system operator; technical conditions of the distribution system operator; URSO decisions; and URSO Decree no. 2/2008 and its amendments[7, 8].

3. Biomass as a renewable source of energy

Biomass is one of the key renewable sources of energy that is produced from organic matter. It includes wood, agricultural crops and waste, and other "living" materials that can be used to produce heat and energy.

Dzurenda and Jandačka [12] define biomass as a matter of biological origin, which includes plant biomass grown in soil and water, animal biomass, production of animal origin, and organic waste.

Directive 2001/77/EC [13] defines the biomass as a biodegradable fraction of products, waste, and residues from agriculture (including vegetal and animal substances), forestry, and related industries, as well as the biodegradable fraction of industrial and municipal waste.

Lieskovský and Gejdoš [14] understand the term biomass as all living and organic matter in each system that was originated and developed as a product of life processes (development, growth, and reproduction) of living organisms. According to this definition, it provides a very wide range of its possible systematic sorting and distribution.

In terms of its origin, we can talk about plant biomass (phytomass), animal biomass (zoomass), and municipal and industrial waste. Dendromass is an organic matter of woody and shrubby plants consisting of wood, bark, and green matter [12]. Phytomass is a biomass of plant origin [15].

Regardless of source, biomass materials can be divided into two broad categories: woody and non-woody. Forests provide only woody materials; agriculture sources provide both woody and non-woody biomass for bioenergy production [16].

The choice of biomass (i.e., woody or herbaceous species) for energy production purposes depends upon the end-use, bio-conversion portion of interest, e.g., combustion, gasification, pyrolysis, fermentation, or mechanical extraction of oils.

Looking back at the recent past, also in the Slovak Republic, biomass for energy purposes was not an interesting topic until 2000. Traditionally, it was previously considered as an additional source of energy to meet local heating needs, mostly in areas without fossil fuel infrastructure. Until 1999, there was no domestic demand for forest fuel chips and their annual production ranged from 2000 to 3000 tons (i.e., 2–3 Kt) [17].

The pioneer in this area was Slovenské energetické strojárne (SES), a. s., Tlmače in Slovakia, which reconstructed the boiler room in 2001 and adapted the equipment for the combustion of chips. According to the TREND newspaper (published on April 11, 2003), at that time it was 20,000 tons of wood chips per year, covering the heating needs of buildings and part of the Lipník housing estate in Tlmače. The use of wood chips in SES Tlmače also solved to a large extent the problem of the Forest Enterprise Levice (LZ), who were looking for sales opportunities for not very attractive tree species such as Turkey oak (*Quercus cerris*) and black locust (*Robinia pseudoacacia*) at that time.

Since that time, much has changed in the timber market. The amount of logging in the Slovak forests has been increasing in the past 15 years. Planned and actual logging is increasing in Slovakia, particularly due to an increase in the share of stands of higher (ruby) age.

The unbalanced age structure in the forests of Slovakia causes cyclical changes also in the development of logging possibilities. It is anticipated that they will decline already around 2030 but depending on the extent of incidental felling [18].

In 2000, approximately 5.5 mil. m^3 of timber was logged, while it was more than 9.3 mil. m^3 in 2017. The trend of the decreasing quality of timber on the market and an increase in the share, especially of the fifth-grade timber assortments, is visible. To a large extent, wood degradation is also due to a high proportion of incidental felling, which regularly exceeds 50% (57% in 2015), with a significant proportion of felling being found in coniferous forests [19].

According to the document "Utilization of wood for energy purposes," the total consumption of solid wood fuel biomass (fuel wood, chips, fine-grained and lump residues after processing and handling of wood, briquettes, and pellets) amounted to 3.05 mil. tons (3.05 MMt) in 2017.

The key consumers of wood fuels, which are the dominant renewable energy source in Slovakia, are the wood processing and pulp and paper industry, the population, central heating sources, and the energy sector. The heat produced i s mainly used for heating and industrial purposes. The proportion of wood fuels in the total consumption of primary energy sources in the Slovak Republic was 1.9% [18].

The heat producers associated in the Slovak Association of Heat Producers (SZVT) heat 38 places together, for which approximately 257,000 tons of timber are used annually (i.e., 2.14% of harvested wood). If the heat producers only used branches and wood waste for heating, it would still not be even 10% of the total harvested wood plant.

Other nine electricity producers from biomass, who are not associated with the SZVT, utilize approximately 530,000 tons (530 Kt) of wood annually, i.e., approximately 4.17% of total timber harvesting. Indication of how much wood is used for individual heating of households is not available [20].

The decisive legal document for forest management in the Slovak Republic is the Act no. 326/2005 Coll. [21] on forests, as amended. The Act defines the areas of forest land and forest protection, professional and differentiated forest management, forest use, and the requirement of sustainable forest management.

The current forestry and agriculture legislation also addresses land use issues related to the sustainability of forest biomass (also dendromass) production and has a direct impact on its energy use [22].

Table 2 presents data representing the development in the dendromass stock specified for energy use.

The expected significant increase in the proportion of renewable energy sources and the use of underproductive agricultural land for the cultivation of energy stands results also in a significant increase in the potential of energy-efficient biomass to produce heat and energy in Slovakia. At the same time, it is possible to support further development of the fuel dendromass market. The amendment to the Act on forests introduced concepts such as energy stands and forest plantations. Energy stands are purpose-built forests with the aim of maximizing biomass production in the first 15 years, while also fulfilling other forest functions, especially soil conservation, erosion control, and partly landscape creation. Biomass produced in this way should be used mainly for energy production.

In energy stands and forest plantations, it is not possible to effectively use the management methods as in conventional forests. For example, it is unreasonable to require the provision of conventional management operations in such forest stand. For that reason, the application of the conventional stand management obligation is excluded in these cases. At the user's request, the stands can be reclassified to

Year	Forest chips[1]		Wood fuel and other[2]		Total	
	(Kt)	(TJ)	(Kt)	(TJ)	(Kt)	TJ
2017	580	5510	845	8028	1425	13,538
2016	610	5795	830	7885	1440	13,680
2015	615	5843	835	7933	1450	13,775
2013	620	5890	820	7790	1440	13,680
2010	250	2375	695	6602	945	8977
2005	120	1140	640	6080	760	7220
2000	5	48	471	4475	476	4522
1990	2	19	368	3496	370	3515

[1]*Wood chips and wood to produce wood chips*
[2]*Fuel wood and wood used for energy from waste, harvest residues, and dead trees*

Table 2.
Development of the dendromass stock for energy use (Source: NLC 1991–2018).

energy stands during the recovery of the Forest Management Program (PSL). In 2006, almost 550 ha of forest were reclassified this way in the OZ Levice (management unit of Forests Slovakia, S.E.). These were mostly the coppices of black locust (96.1%) and Turkey oak (1.2%). These coppices are restored by the clear cutting connected with the maximum utilization of the stump and root sprouting of the abovementioned tree species [23].

Current resources of wood on non-forest ground are mainly the tree stands on long-term unused agricultural land (so-called white areas), streamside stands, and trees in the open country, including linear planting vegetation, e.g., windbreaks and trees around roads.

Legislative conditions for planting fast-growing trees on agricultural land are determined directly by the Act no. 220/2004 Coll. [24] on the protection and use of agricultural land and by the amendment of Act no. 245/2003 Coll. on integrated pollution prevention and control and on amendments and supplements to certain acts.

For the purposes of this Act, fast-growing trees on agricultural land shall mean the plantation of fast-growing trees to produce wood biomass, on an area with extent more than 1000 m^2, for a maximum of 20 years.

The fast-growing tree species can be planted on agricultural land classified into the 5th to 9th quality group, according to the code of a certified soil-ecological unit used in Slovakia. Also they can be planted on agricultural land contaminated by dangerous substances, or on agricultural land classified into the 3rd or 4th quality group according to the code of a certified soil-ecological unit, or on agricultural land, which is located in a floodplain, that is wet or exposed to wind erosion. The plantations of fast-growing tree species cannot be established on areas situated in the 3rd to 5th degree of nature and landscape territorial preservation.

The tree stands on "white areas" formed mainly by succession of trees are located on an area of ca. 275,000 ha with a total wood supply of 36.6 MMm^3 (timber stock without bark).

The current stock of coniferous trees is 12.7 MMm^3, hard deciduous trees 9.1 MMm^3, and soft deciduous trees 14.8 MMm^3. The assortment structure of stands on "white areas" is represented by a higher proportion of fiber wood and wood for energy use than the stands on forest land. Due to their localization, stands on "white areas" are easily accessible, and terrain conditions enable the use of efficient timber logging technologies [25].

Another possibility of increasing biomass production is the plantation of fast-growing trees. The establishment of fast-growing tree plantations supports other unique and important environmental and ecological benefits that can provide enough raw material for the energy industry. At the same time, if certain decisions are considered in addition to production when planning a fast-growing tree plantation, they finally can have a positive impact on the landscape, biodiversity, soil, and water cycle in the ecosystem. The use of this method of targeted energy biomass extraction is a combination of forestry and agriculture and brings new opportunities supporting regional energy self-sufficiency.

With the increasing demands for biomass for energy purposes, the issues of production and targeted cultivation of fast-growing tree plantation (known also as short rotation coppice (SRC)) are becoming topical. In the future, demand for wood as a raw material for heating and electricity production is expected to increase. This increase will mainly be influenced by the situation on the fuel market and will be supported as a target of national and European energy policy. Energy chips from fast-growing tree species can thus make a significant contribution to the European targets related to increasing the proportion of renewable energy sources [23].

The most frequently planted tree species on plantations are various clones and varieties of poplar (*Populus* sp.) and willow (*Salix* sp.). Current legislation does not directly limit plantation owners and users to the use of a clone or varieties, but the cultivation of non-origin tree species is in violation of the Act no. 543/2002 Coll. [26] on nature and landscape. Appropriate selection considering habitat conditions is a prerequisite for meeting production expectations.

An important factor that can influence the future plantation of fast-growing trees is enough potential area for their establishment. The potential of plantation establishment is both on the forest and in the agricultural ground fund.

In 2017, the area of utilized agricultural land was 1,910,654 ha. The Slovak Republic accounts for 38.8% of agricultural land in the total land area [18]. In addition, the distribution of agricultural land in the Slovak Republic is also characterized by a high proportion of agricultural land in mountain and foothill areas with rugged terrain and unfavorable climatic conditions.

Under such conditions, intensive agricultural production is not efficient today. However, it creates the preconditions for the possibility of diversification of production, one alternative of which is the production of biomass for energy purposes.

In the medium-term horizon, energy stands is considered to be planted on an area of 30,000 ha. Their production of energy chips is accounted for 70% and fiber wood for 30 %, considering the 15-year-long rotation period (MP SR 2018) [18].

The possibilities of biomass to be used for energy purposes and its energy properties are studied by many experts worldwide. There is introduced brief review of the last research works in this field.

The worldwide research trends related to biomass as renewable energy derived from the analysis of the state of the research and trends in biomass for renewable energy from 1978 to 2018 were published by Perea-Moreno et al. [27]. Woch et al. [28] published a case study focusing evaluation of potential use of forest biomass for renewable energy based on systems approach. The ways to meet the future energy demands based on biotechnology and wood for energy purposes are described by Al-Ahmad [29]. Climate, economic, and environmental impacts of producing wood for bioenergy are introduced by Birdsey et al. [30]. Koponen et al. [31] published a study in which they tried to quantify the climate effects of bioenergy. Cordiner et al. [32] introduced results of biomass pyrolysis modeling at laboratory scale, which were further completed with their experimental validation. Kluts et al. [33] dealt with agriculture biomass sources. There are also several studies focusing on the determination of energetic parameters of biomass, e.g., [34–37].

4. Assessment of woody biomass stock in Slovak forests: case study 1

The geodatabase containing the data from the territory of the Slovak Republic (digital terrain model (DTM), settlements, district borders—producer and provider is the Topography Institute of the col. Jan Lipsky in Banska Bystrica) was added and preprocessed in ArcGIS for Desktop ver. 10.2. together with geodata on forest stand outlines and forest inventory database produced and provided by the Department of Forest Resources and Informatics of the National Forest Centre in Zvolen and containing the detailed description of the stands which is updated every 10 years.

For the needs of further analyses of the database, all the forest stands existing in the territory of the Slovak Republic were selected. Totally, there were 211,968 forest stands included in the analysis.

From the digital terrain model, and using the surface analyses tools in ArcGIS, a raster of terrain slopes in the ArcGIS environment, which was later used in the process of identifying available sources of woody biomass (dendromass) for energy use in forests of the Slovak Republic, was derived.

As the primary source of data for calculating the amount of dendromass is available, we used the data concerning the description of the basic parameters of forest stands, which are introduced in the database, which is used as the primary source of data for providing the spatial analyses in GIS environment. These data are the result of detailed surveys on forest which are provided for purposes of forest management plan elaboration.

As the basic parameters for the derivation of the total available dendromass stock, we used the data on the extent of area of forest land, timber stock, and the planned annual cutting. In addition, we also analyzed the age structure and forest category of the stands.

Not all dendromass is suitable for energy purposes. There were specified restriction criteria. The most restrictive criterion to identify the dendromass for energy purposes is terrain slope. Steep terrain is a limitation for deployment of majority of timber logging technologies used in Slovakia. That is the reason why the forest stands situated within terrain with slope of 50% and more were excluded.

There were also excluded forest stands classified as protection forests, where protection function is superior to productive function. There are also included stands assigned into the 5th degree (the highest) of nature preservation, which are mostly in the National Parks of the Slovak Republic.

The information on the category of forest and its nature protection level was obtained from a database containing basic parameters of forests, which we received from the Department of Forest Resources and Informatics, National Forest Centre. Those data were classified, and the unsuitable stands were excluded from further analyses.

Another criterion for excluding the stands unsuitable for energy purposes was the classification code of individual forest stands related to the "management set of forest types," which is used in the Slovak Republic. The management sets of forest types [38] were identified, which are naturally very low in nutrients (especially calcium, magnesium and potassium) or habitats with extreme texture, skeleton, water regime, as well as sites with an excess of certain nutrients, but a great lack of potassium and phosphorus. The forest stands belonging to those management sets of forest types were classified as unsuitable and were excluded from the analysis.

The results of the analyses are introduced in **Figure 5** and in **Table 3**. The information on potential woody biomass stock was derived from eight existing district (region) units in Slovakia.

The graphical output of the analysis is introduced in **Figure 5**.

In **Table 3**, the potential stock of woody biomass to be used for energy purposes in Slovakia, derived from eight existing districts in Slovakia based on pre-defined criteria, is introduced.

According to the data introduced in **Table 3**, we can state that the highest stock of woody biomass to be potentially used for energy purposes is in Banská Bystrica (24%) and Prešov (24%) districts (together 48% of the overall woody biomass stock).

Further, we introduce the approaches used to determine the energetic properties of selected fast-growing tree species which have potential to be planted for energy purposes in Slovakia.

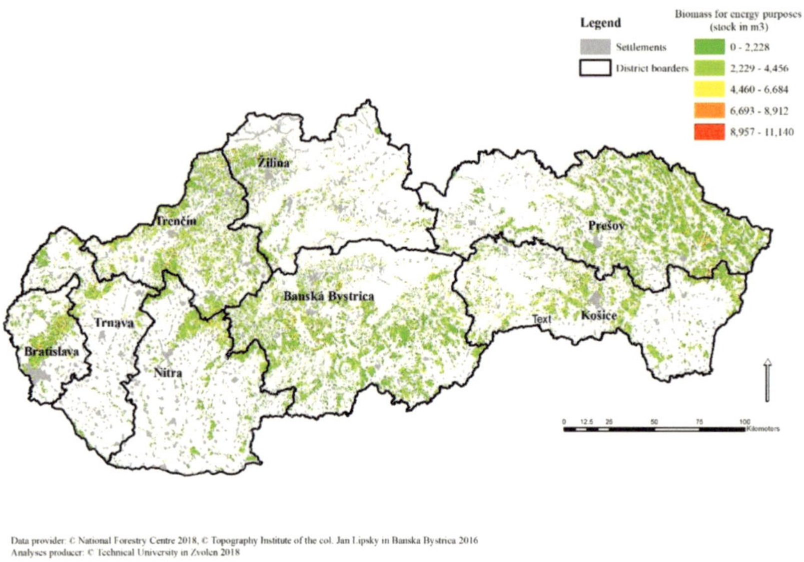

Figure 5.
Woody biomass stock in forests of the Slovak Republic (Source: Authors).

District	Biomass stock (m^3)	Number of forest stands	Stand extent (km^2)	District extent (km^2)
Bratislava	10,782,036	9955	418.0	2059.5
Trnava	10,034,628	11,109	414.3	4143.0
Trenčín	37,080,538	32,030	1336.8	4501.3
Žilina	16,716,540	19,726	590.0	6789.2
Nitra	15,183,661	15,200	732.2	6338.7
Banská Bystrica	53,848,038	54,345	2292.3	9450.5
Košice	25,116,195	19,536	1041.3	6749.2
Prešov	53,896,725	46,659	2307.5	8988.2
Total	222,658,361	208,560	9132	49,020

Table 3.
Woody biomass stock to be used for energy purposes in Slovakia.

5. Fire and energy properties of woody biomass: case study 2

To analyze the fire and energetic properties of selected species of woody biomass for energy production purposes, several standardized but also progressive analytical methods were used.

Three woody biomass species were tested: *Populus x euroamericana* clone MAX 4, *Salix viminalis* clone TORA, and *Paulownia tomentosa*.

To implement the laboratory fire tests, the samples of woody biomass species were represented by the blocks with dimensions of 50 × 40 × 20 mm in the case of mass loss testing and 20 × 20 × 10 mm in the case of spontaneous ignition temperature testing.

The samples of *Salix viminalis* clone TORA and *Populus x euroamericana* clone MAX 4 were taken from the existing plantations of the University Forest Enterprise of the Technical University in Zvolen territory.

The samples of *Paulownia tomentosa* were taken from the plantations belonging to the Agricultural Co-operative Dolné Saliby.

The following analyses were implemented: analyses of spontaneous ignition temperature; analyses of mass loss during sample thermal loading with radiant heat source; and gross calorific value, heating value, and ash content analyses.

5.1 Analyses of spontaneous ignition temperature and induction period of samples

In the laboratory fire tests, samples of woody species, i.e., blocks with dimensions of 50 × 40 × 20 mm for mass loss testing and 20 × 20 × 10 mm for spontaneous ignition temperature testing, were used.

Before the test, all the samples were conditioned according to the STN EN ISO 291 standard requirements. Totally, three samples of each woody biomass and herbaceous energy crops undergone testing.

To determine the temperature of spontaneous ignition, the incendiary hot-air oven (Setchkin furnace) was used, and the methodology for testing the spontaneous ignition temperature, according to the STN ISO 871 standard, was applied.

Those analyses were performed in the laboratories and use the research infrastructure of the Department of Fire Protection, Faculty of Wood Sciences and Technology, Technical University in Zvolen.

Table 4 shows an overview of the determined spontaneous ignition temperatures and induction periods reached by *Populus x euroamericana* clone MAX 4.

The lowest mean spontaneous ignition temperature value was recorded by *Salix viminalis* clone TORA (419.46°C), which was reached in 328.87 s from the start of the test. The results also showed that with increasing thermal loading (and higher spontaneous ignition temperature value), the samples were resistant to fire for a shorter time.

5.2 Analysis of mass loss during sample thermal loading with radiant heat source

To understanding the thermal decomposition process of all the samples tested during their burning, implementing thermal analyses, and studying the mass loss of the sample are recommended.

To study the mass loss of the samples, the nonstandard method of solid thermal properties testing was applied.

The samples of woody biomass and energy crops undergone thermal loading by a radiant heater with the power of 1000 W for a specific time, i.e., 10 min. The mass

Measurement no.	Spontaneous ignition temperature of *Populus* *t* (°C)	Spontaneous ignition temperature of *Salix* *t* (°C)	Spontaneous ignition temperature of *Paulownia* *t* (°C)
1.	424.92	412.65	420.10
2.	417.14	426.63	410.98
3.	419.69	419.09	441.87
Mean	420.58	419.46	424.32

Table 4.
Spontaneous ignition temperatures of woody biomass species.

loss of the samples (g) was measured for each 10-s interval. Totally, three samples of each woody biomass undergone testing.

Those analyses were performed in the laboratories and use the research infrastructure of the Department of Fire Protection, Faculty of Wood Sciences and Technology, Technical University in Zvolen.

The resulting courses of mass loss of the tested woody biomass species are introduced in **Figures 6–8**.

5.3 Gross calorific value, heating value, and ash content analyses

To calculate the heating value, it was necessary to determine the gross caloric value of the samples. The IKA C200 calorimeter was used to determine it. The procedure was conducted in correspondence with the standard STN ISO 1928:2003-07 Solid fuels. Determination of gross caloric value and calculation of heating value. In the test, the sample is burnt in a calorimetric bomb and filled with oxygen under the pressure of 3–5 MPa.

Based on the mathematical Eq. (1) introduced in the same standard, the heating values (KJ·kg^{-1}) of the samples were further calculated:

$$q_{v,net,m} = \left[q_{v,gr,d} - 206.0 \cdot w\,(H)_d\right] \cdot (1 - 0.01 \cdot M_T) - 23.5 \cdot M_T \qquad (1)$$

where $q_{v,net,m}$—heating value at constant volume and containing with water (kJ·kg^{-1}); $q_{v,gr,d}$—gross calorific value at constant volume without water content (kJ·kg^{-1}); $w(H)_d$—percentage of hydrogen (%); M_T—total water content of the fuel for which conversion is required - relative moisture (%).

In the calculations of heating value, the relative moisture content of 10% was used. Before the testing, the samples were dried at 103 ± 2°C to reach the moisture content of 0% and further conditioned in a desiccator at the temperature of 20 ± 1°C for 24 hrs. Three measurements were made for each sample. The results show the average value of those measurements.

The procedure for ash determination was based on the requirements of the standard STN ISO 1171:2003 (441378) Solid mineral fuels. Determination of ash. The principle of the method is the incineration of the sample, heated in air at a temperature of 815 °C ± 10 °C, for specified time interval and maintained at that above-mentioned constant temperature. For this purpose, the muffle furnace was used. The ash content was calculated from the weight of the residue after incineration.

Those analyses were performed at laboratories and use the research infrastructure of the Department of Forest Harvesting, Logistics and Ameliorations, Faculty of Forestry, Technical University in Zvolen.

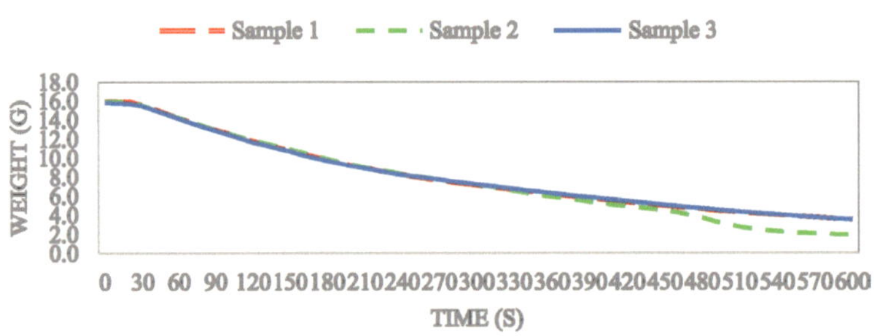

Figure 6.
Mass loss course of Populus x euroamericana *clone MAX 4 during the thermal loading.*

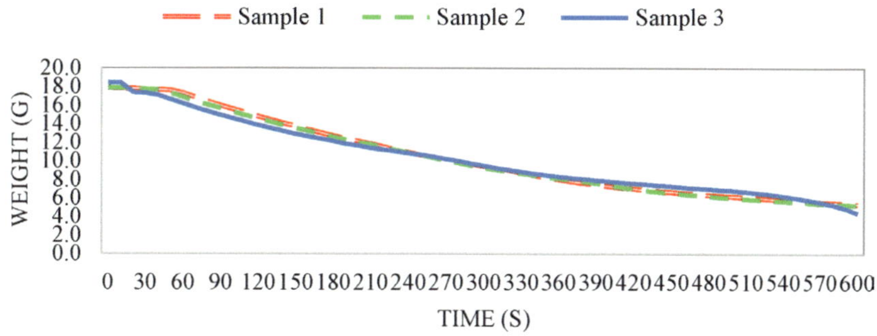

Figure 7.
Mass loss course of Salix viminalis *clone TORA during the thermal loading.*

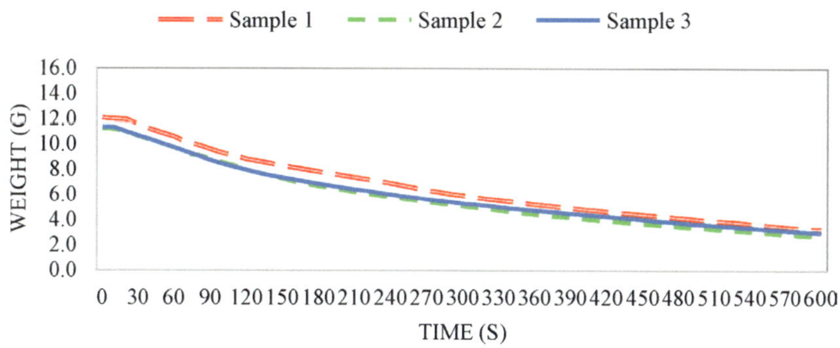

Figure 8
Mass loss course of Paulownia tomentosa *during the thermal loading.*

The highest energy potential expressed in terms of the highest gross calorific values (19.71 ± 0.18 MJ·kg^{-1}) as well as heating values was found in the case of *Paulownia tomentosa*.

The results of gross calorific value and heating value of all woody biomass species are introduced in **Table 5**.

The ash content analyses results are introduced in **Table 6**.

The highest ash content was achieved in *Populus x euroamericana* clone MAX 4, followed by *Salix viminalis* clone Tora and *Paulownia tomentosa* which showed the lowest value of ash content in the analyses.

The highest energy potential expressed in terms of the highest gross calorific values as well as heating values (at 10% moisture content) was recorded in the case of *Paulownia tomentosa* (19.71 ± 0.18 MJ·kg^{-1}; 16.40 ± 0.18 MJ·kg^{-1}). The lowest values of gross calorific values and heating were recorded in *Populus x euroamericana* clone MAX 4 (19.47 ± 0.29 MJ·kg^{-1}; 16.18 ± 0.29 MJ·kg^{-1}). The differences in values recorded in the tested samples of fast-growing tree species were very low. According

Sample	Gross calorific value (MJ·kg^{-1})	Heating value (MJ·kg^{-1})	Standard deviation
Populus x euroamericana clone MAX 4	19.47	16.18	0.29
Salix viminalis clone Tora	19.63	16.33	0.11
Paulownia tomentosa	19.71	16.40	0.18

Table 5.
Gross calorific value and heating value of tested woody biomass samples.

Sample	Ash content (w%)	Standard deviation
Populus x euroamericana clone MAX 4	2.58	0.24
Salix viminalis clone Tora	1.28	0.08
Paulownia tomentosa	0.75	0.05

Table 6.
Ash content of the woody biomass species.

to these finding, all the tested biomass species were considered suitable to be used for further energy use. However, *Paulownia tomentosa* seems to be the most suitable from calorific value and heating value point of view.

Heating value should be tightly connected also with elemental composition and affected by the variation in cell wall composition and ash. This fact was confirmed also by ash content analysis using the muffle furnace for ashing. The ash content of tested woody biomass species was in the range of 0.75–2.58 w%. The lowest values of ash content were recorded right in *Paulownia tomentosa* (0.75 ± 0.05 w%).

Similar results were achieved also by Yavorov et al. [37], who were engaged in determining the potential of fast-growing hardwood species from Bulgaria (*Paulownia elongata*, *Populus alba*, and *Salix viminalis* RUBRA), and Martinka et al. [39] who studied the calorific value and fire risk of selected fast-growing hardwood species (*Populus nigra x Populus maximowiczii*, *Salix alba* L.).

6. Conclusions

Climate change caused by increasing greenhouse gas emissions is among the most serious global threats. Therefore, many experts have been looking for ways to solve this problem for more than 20 years.

In recent years, in the world, the importance of the energy sector has increased, particularly in terms of sustainable development. The direction of energy sector development is slowly changing toward the use of environmentally friendly fuels and energy from renewable sources.

Slovakia as a country that is more than 90% dependent on imports of primary energy sources should have a primary interest to use its own, renewable energy sources.

Biomass is the largest renewable energy source in Slovakia. It consists of vegetable and animal origin materials suitable for energy use. Biomass is considered in terms of CO_2 biomass is a neutral fuel, because it shall release only as much of the CO_2 when burning as the plant has taken during its growing.

The Energy Policy of the Slovak Republic aims in increasing the share of renewable and secondary energy sources, which constitutes a significant portion of woody biomass (dendromass) produced in forestry, wood industry, and pulp and paper industries.

To identify the available sources of woody biomass or any kind of biomass, as the first step of any analysis concerning the possible location of any power plant using biomass for energy and heat production, it is recommended to deploy those tools allow processing the existing data on forest and agricultural land and different kinds of spatial analyses to get the required information. An approach that is used in Slovakia for this purpose is introduced.

To study the important characteristics for the biomass combustion process is possible through deployment of standardized and progressive nonstandardized laboratory fire tests and calorimetric and thermal analyses. Some of them were introduced in the framework of this chapter.

Acknowledgements

This work was supported by the Slovak Research and Development Agency, based on the Agreements no. APVV-17-0005 (20%) and APVV-16-0487 (20%), VEGA Grant Agency under project VEGA 1/0493/18 (20%), KEGA Grant Agency under projects KEGA 032PU-4/2018 (20%) and MPRVSR item 08V0301 – Research and development to promote forestry competitiveness (SLOV – LES) and OPVaI International Teaming Research Centers (20%).

Nomenclature

EU	European Union
RES	Renewable Energy Sources
URSO	Regulatory Office for Network Industries
MP SR	Ministry of Agriculture of the Slovak Republic
EC	European Commission
CHP	Combined Heat and Power
GDP	Gross Domestic Product
SR	Slovak Republic
DTM	Digital Terrain Model

Symbols

$q_{v,net,m}$	Heating value at constant volume and containing with water [kJ·kg^{-1}]
$q_{v,gr,d}$	Gross calorific value at constant volume without water content [kJ·kg^{-1}]
$w(H)_d$	Percentage of hydrogen [%]
t	Temperature [°C]

Author details

Andrea Majlingová[1*], Martin Lieskovský[2], Maroš Sedliak[3] and Marián Slamka[3]

1 Faculty of Wood Sciences and Technology, Department of Fire Protection, Technical University in Zvolen, Zvolen, Slovakia

2 Faculty of Forestry, Department of Forest Harvesting, Logistics and Amelioration, Technical University in Zvolen, Zvolen, Slovakia

3 National Forest Centre, Forest Research Institute, Zvolen, Slovakia

*Address all correspondence to: majlingova@tuzvo.sk

References

[1] Eurostat. Renewable Energy Statistics [Online]. 2019. Available from: https://ec.europa.eu [cited 03 June 2019]

[2] Directive 2009/28/EC of the European Parliament and of the Council of 23 April 2009 On the Promotion of the Use of Energy from Renewable Sources and Amending and Subsequently Repealing Directives 2001/77/EC and 2003/30/EC. Available from: https://eur-lex.europa.eu

[3] Regulation (EU) 2018/1999 of the European Parliament and of the Council on the Governance of the Energy Union and Climate Action, Amending Regulations (EC) No. 663/2009 and (EC) No. 715/2009 of the European Parliament and of the Council, Directives 94/22/EC, 98/70/EC, 2009/31/EC, 2009/73/EC, 2010/31/EU, 2012/27/EU and 2013/30/EU of the European Parliament and of the Council, Council Directives 2009/119/EC and (EU) 2015/652 and Repealing Regulation (EU) No 525/2013 of the European Parliament and of the Council [Online]. Available from: https://eur-lex.europa.eu [cited 17 February 2020]

[4] Directive 2012/27/EU of the European Parliament and of the Council of 25 October 2012 On Energy Efficiency, Amending Directives 2009/125/EC and 2010/30/EU and Repealing Directives 2004/8/EC and 2006/32/EC Text with EEA Relevance. Available from: https://eur-lex.europa.eu

[5] European Parliament News. Energy: New Target of 32% from Renewables by 2030 Agreed by MEPs and Ministers [Online]. 2018. Available from: https://www.europarl.europa.eu/news/en/press-room/20180614IPR05810/energy-new-target-of-32-from-renewables-by-2030-agreed-by-meps-and-ministers [cited 17 February 2020]

[6] Energie Portál. Share of Energy from Renewable Sources in Final Energy Consumption in SR in Period 2004-2016 [Online]. 2019. Available from: https://www.energie-portal.sk/ [cited 26 November 2018]

[7] Decree of the Regulatory Office for Network Industries Laying Down Details of the Request and the List of Documents for Exemption from the Obligation to Provide Third Party Access to Networks and Storage Capacity for New Significant Gas Facility or Refurbished Gas Facility. Available from: http://www.urso.gov.sk

[8] URSO. Annual Report 2018 [Online]. 2019. Available from: http://www.urso.gov.sk [cited 11 April 2019]

[9] Act of the National Council of the Slovak Republic No. 656/2004 Coll. "Energy Act". Available from: https://www.zakonypreludi.sk/

[10] Regulation of the Government of the Slovak Republic No. 211/2010, Amending and Supplementing Regulation of the Government of the Slovak Republic No. 317/2007 Coll. Laying Down the Rules for the Functioning of the Electricity Market, as Amended by Act No. 309/2009 Coll [Online]. Available from: https://www.slov-lex.sk [cited 17 November 2019]

[11] Act No. 309/2009 Coll. on the Promotion of Renewable Energy Sources and High Efficiency Cogeneration and on Amendments to Certain Acts. Available from: https://www.zakonypreludi.sk

[12] Dzurenda L, Jandačka J. Energetické využitie Biomasy [Energy Use of Biomass]. Zvolen: Technical University in Zvolen; 2010. p. 161

[13] Directive 2001/77/EC of the European Parliament and of the Council of 27 September 2001 On the Promotion of Electricity Produced

from Renewable Energy Sources in the Internal Electricity Market. Available from: https://eur-lex.europa.eu

[14] Lieskovský M, Gejdoš M. Komplexné využitie Biomasy v Lesnom hospodárstve [Complex Use of Biomass in Forestry]. Zvolen: Technická univerzita vo Zvolene; 2016. p. 136

[15] Hutňan M. Využitie fytomasy ako obnoviteľného zdroja na energetické účely [Use of phytomass as a renewable source for energy purposes]. In: Pospíšil R, editor. Využitie Biomasy z Obnoviteľných Zdrojov na Energetické Účely. Nitra: Slovenská Poľnohospodárska Univerzita v Nitre; 2012. pp. 163-170

[16] Jawaid M, Paridah M, Tahir PM, Saba N. Lignocellulosic Fibre and Biomass-Based Composite Materials—Processing, Properties and Applications. London, UK: Woodhead Publishing; 2017. p. 522

[17] Oravec M. BIOCLUS—Developing Research and Innovation Environment in five European Regions in the field of Sustainable Use of Biomass Sources (Project 245438)—Regionálna Stratégia Slovenska 2010 [Online]. 2010. Available from: http://www.bioclus.eu [cited 02 December 2018]

[18] MP SR. Green Report 2017 [Online]. 2018. Available from: http://www.mpsr.sk/en/index.php?navID=16&id=68 [cited 21 March 2019]

[19] Gejdoš M, Lieskovský M. Aké Hrozia biologické riziká Pri Dlhodobom skladovaní lesných štiepok? [What Are the Biological Risks of Long-Term Storage of Forest Chips?]. In: Správy z Výskumu Lesníckej Fakulty Pre Prax. Zvolen: Vydavateľstvo TU vo Zvolene; 2017. pp. 39-44

[20] Oravec M, Slamka M. Výskum Potenciálu Drevnej Biomasy na Energetické Využitie v Podmienkach Slovenska [Research of the Potential of Wood Biomass for Energy Use in Slovakia] [Online]. Zvolen: Národné Lesnícke Centrum—Lesnícky Výskumný Ústav Zvolen; 2018. Available from: http://www.nlcsk.sk [cited 02 December 2018]

[21] Act No. 326/2005 Coll. on Forests. Available from: https://www.zakonypreludi.sk

[22] KPMG. Kritériá Udržateľného Využívania Biomasy v Regiónoch Slovenska pre Programy SR na Obdobie 2014-2020 Spolufinancované z EŠIF-So Zameraním na Drevnú Biomasu [Criteria for Sustainable Use of Biomass in Regions of Slovakia for 2014-2020 Programs of the SR Co-financed by the European Structural Investment Fund-Focusing on Wood Biomass] [Online]. 2016. Available from: https://www.op-kzp.sk [cited 02 December 2018]

[23] Messingerová V, Lieskovský M, Gejdoš M, Slugeň J, Mokroš M, Tomaštík J. Technika a Technologické Postupy Pri Produkcii Biomasy a Jej Energetickom Zhodnotení [Technology and Technological Procedures in Biomass Production and its Energy Recovery]. Zvolen, Slovakia: Technická Univerzita vo Zvolene; 2016. p. 101

[24] Act No. 220/2004 Coll. on the Protection and Use of Agricultural Land and on the Amendment of Act No. 245/2003 Coll. on Integrated Pollution Prevention and Control and on Amendments and Supplements to Certain Acts. Available from: https://www.zakonypreludi.sk

[25] Oravec M, Bartko M, Slamka M. Postupy Intenzifikácie Produkcie Drevnej Biomasy na Energetické Využitie [Procedures for Intensifying Wood Biomass Production for Energy Use]. Zvolen, Slovakia: Národné Lesnícke Centrum—Lesnícky Výskumný Ústav; 2012. p. 65

[26] Act No. 543/2002 Coll. on Nature and Landscape Protection. Available from: https://www.zakonypreludi.sk

[27] Perea-Moreno MA, Sameron-Manzano E, Pera-Moreno AJ. Biomass as renewable energy: Worldwide research trends. Sustainability. 2019;**11**:863-881

[28] Woch F, Hernik J, Sankowski E, Pioro P, et al. Evaluating the potential use of forest biomass for renewable energy: A case study with elements of a systems approach. Polish Journal of Environmental Studies. 2020;**29**(1):885-891

[29] Al-Ahmad H. Biotechnology for bioenergy dedicated trees: Meeting future energy demands. Zeitschrift Fur Naturforschung, Section C-A Journal of Biosciences. 2018;**73**(1-2):15-32

[30] Birdsey R, Duffy P, Smyth C, Akurz W, et al. Climate, economic, and environmental impacts of producing wood for bioenergy. Environmental Research Letters. 2018;**13**(5)

[31] Koponen K, Soimakallio S, Kline KL, Cowie A, et al. Quantifying the climate effects of bioenergy—Choice of reference system. Renewable & Sustainable Energy Reviews. 2018;**81**:2271-2280

[32] Cordiner S, Manni A, Mulone V, Rocco V. Biomass pyrolysis modeling of systems at laboratory scale with experimental validation. International Journal of Numerical Methods for Heat & Fluid Flow. 2018;**28**(2):413-438

[33] Kluts IN, Brinkman MLJ, De Jong SA, Junginger HM. Biomass resources: Agriculture. Biorefineries. 2019;**166**:13-26

[34] Liang F, Zhang T, Xiang H, et al. Pyrolysis characteristics of cellulose derived from moso bamboo and poplar. Journal of Thermal Analysis and Calorimetry. 2018;**132**:1359-1365

[35] Majlingová A, Zachar M, Lieskovský M, et al. The analysis of mass loss and activation energy of selected fast-growing tree species and energy crops using the Arrhenius equation. Acta Facultatis Xylologiae Zvolen. 2018;**60**(2):177-188

[36] Rodrigues A, Nunes LJR, Godina R, Matias JCO, et al. Correlation between chemical alterations and energetic properties in Torrefied biomass. In: 2018 IEEE International Conference on Environment and Electrical Engineering and 2018 IEEE Industrial and Commercial Power Systems Europe (EEEIC/I&CPS Europe); 2018

[37] Yavorov N, Petrin ST, Valchev I, et al. Potential of fast-growing poplar, willow and Paulownia for bioenergy production. Bulgarian Chemical Communications. 2015;**47**(SI:A):5-9

[38] Ilavský J et al. Štúdia o Dostupných Zdrojoch Biomasy a Ich Efektívnom Zabezpečení na Výrobu Energie Vo Zvolenskej Teplárenskej a.s. [Study on Available Biomass Resources and their Effective Ensuring for Power Generation in Zvolen Heat Plant, Monograph]. 1st ed. Zvolen: Technical University in Zvolen; 2006

[39] Martinka J, Martinka F, Rantuch P, Hrušovský I, Blinová L, Balog K. Calorific value and fire risk of selected fast/growing wood species. Journal of Thermal Analysis and Calorimetry. 2018;**131**:899-906